THE LONG SHADOW
OF THE PARAFINITE

Three Scenes from the Prehistory of a Concept

O. Bradley Bassler

Docent
Press

DOCENT PRESS
Boston, Massachusetts, USA
www.docentpress.com

Docent Press publishes monographs and translations in the history of mathematics for thoughtful reading by professionals, amateurs and the public.

O. Bradley Bassler studied in the Committee on Social Thought at the University of Chicago and took a second Ph.D. in Mathematics at Wesleyan University. He is Associate Professor of Philosophy at the University of Georgia, Athens, USA and is the author of *The Pace of Modernity: Reading with Blumenberg* (re.press, 2012).

Cover design by Nathaniel Parker Raymond, Graphic Details, Portsmouth, New Hampshire.

Contents

for my mother, my first and best teacher,

who answered all the questions that mattered most

Abbreviations

CDF	J. L. McCauley, *Chaos, Dynamics and Fractals: an algorithmic approach to deterministic chaos*
DLC	H. J. M. Bos, "Differentials, Higher-Order Differentials and the Derivative in the Leibnizian Calculus"
DSR	Leibniz, *De Summa Rerum: Metaphysical Papers, 1675–1676*
GO	Galileo Galilei, *Opere di Galileo Galilei* (volume 8)
GMT	Jacob Klein, *Greek Mathematical Thought and the Origins of Algebra*
LP	Reviel Netz, *Ludic Proof: Greek Mathematics and the Alexandrian Aesthetic*
MS	Leibniz, *Mathematische Schriften*
NE	Leibniz, *New Essays on Human Understanding*
PE	Leibniz, *Philosophical Essays*

PPL   Leibniz, *Philosophical Papers and Letters*

PR   Wittgenstein, *Philosophical Remarks*

PS   Leibniz, *Philosophische Schriften*

SD   Reviel Netz, *The Shaping of Deduction in Greek Mathematics: A Study in Cognitive History*

STKC   J. Alberto Coffa, *The Semantic Tradition from Kant to Carnap: To the Vienna Station*

TLP   Wittgenstein, *Tractatus Logico-Philosophicus*

TM   Reviel Netz, *The Transformation of Mathematics in the Early Mediterranean World: From Problems to Equations*

WA   Wittgenstein, *Wiener Ausgabe*

WFFM   Mathieu Marion, *Wittgenstein, Finitism, and the Foundations of Mathematics*

WML   David Stern, *Wittgenstein on Mind and Language*

Preface: Another Book for Everyone and No One

"Can it be seen that each of us is everyone and no one?"

– Stanley Cavell,
A Pitch of Philosophy

Although this volume focuses on questions in the history and philosophy of mathematics, I intend the work offered here as part of a long-term project rethinking the status of metaphysics in the modern European-American tradition.

Socratic wisdom teaches that the unexamined life is not worth living. We all (almost all) of us own radios, televisions, DVD players, cell phones and other "appliances." While real expenditure on food and clothes has decreased as a fraction of total income over the last half-century, percentage of income spent on electronic devices has skyrocketed. We none (almost none) of us know all that much about how they work: we haven't examined them. But that is not what Socrates meant: Socrates meant for you to examine, to know, yourself. This will be a solace for philosophical "types," which is to say, for those of us who wish to know ourselves.

The kicker is that the solace extends only so far, for we cannot know ourselves without knowing how we think. And in a world of proliferating technology it becomes increasingly challenging to deny that more and more of our thinking is taken up with *counting*.

We can live the good life in ignorance that Marconi invented the wireless, but the place of numbers in your life is different from the place the radio occupies that sits on the dashboard of your car: numbers are a basic part of your mental equipment. Numbers are not *an* appliance: they are *the applier*. And we don't understand how *they* work, so we don't understand how *we* do either. Apparently, Socrates wasn't as concerned about this as he should have been.

In this current age of "philosophical specialization," in which each member of the philosophical community, upon entry into the "profession," is required to list AOS and AOC (the need for a translation of these terms distinguishing the outside from the inside), the sort of work I propose here is double trouble: it needs to delve into the "specialities" of the foundations of mathematics, and yet ultimately I pursue these issues only as means towards another end, and one it is largely thought none of the business of "philosophers of mathematics" to consider.

On the other hand, if philosophy is to be thought in the traditional sense, as "love of knowledge," or even as the accession to wisdom, then we are saddled with, indeed strapped by, a set of demands which it has persistently been claimed that the modern worldview, and the modern metaphysical tradition in particular, is unable to sustain. The consequence is that many of the philosophical themes which will prove focal to the larger enterprise this work initiates have been relegated to the antiquarian speciality of "history of philosophy" (my original Area Of Specialization) and to the history of ancient philosophy in particular, with its praxical "renascences" among the philosophical communitarians. Instead of knowing ourselves, we have hired specialists to know about Socrates.

Communitarians or anticommunitarians, philosophical antimodernists or modernists, howsoever: we all live in the present, much as we attempt in various ways to deny it. In this wayward work, I begin the attempt to articulate what might be involved in the possibility of a *contemporary* philosophical orientation, which is to say: one that may look at the present with open eyes, listen with open ears, recognize what is given to us or inflicted upon us, and what is subject to our modification. The investigation of the parafinite helps with this need for orientation by providing a first laboratory in which we might confront basic conceptual problems of philosophical contemporaneity.

In this sense, I intend this work as a book for everyone, and it is in the "merely technical" difficulties, not to be underestimated in our technological age, that the work presents obstacles, is "stood against." It begins to compile an assemblage of tools for modern philosophical orientation. In order to be open to as wide a readership as possible, I have devised various tracks through the book in the "Navigation" guide below. Two other volumes, *The Pace of Modernity: Reading With Blumenberg* (re.press, 2012), and *Diagnosing Contemporary Philosophy With the Matrix Movies* (under contract, Palgrave/MacMillan) also belong to this ongoing venture. A third, *Kant, Shelley and the Visionary Critique of Metaphysics*, awaits its destined publisher.

I begin by invoking three passages from the last hundred years of European philosophy which will serve as a kind of frame for this enterprise. First, a passage from Heidegger, from a lecture series closely associated with his second masterwork, *Contributions to Philosophy*:

> *What would remain of the whole business of archives and literature, what would remain of the business of reviews and dissertations, if by one stroke what is unessential became ungraspable? But that will not happen, for the unessential, in very different forms, is the long shadow cast by the essential, to end up mostly overshadowed by it.*[1]

---

[1]Martin Heidegger, *Basic Questions of Philosophy: Selected "Problems" of*

Heidegger suggests the picture of the long shadow which I will use to track the prehistory of the parafinite in this volume. If the parafinite is, as I indeed think, the "essence" of counting, then we may begin to chart the flow of this essence through the history of mathematics by looking at the long shadow it casts.

Second, I invoke a brief passage from Gilles Deleuze's masterwork, *Difference and Repetition:*

> *Contradictions are not 'resolved', they are dissipated by capturing the problem of which they reflect only the shadow.*[2]

I will attempt to show that the prehistory of the parafinite is a history of *impasse*. Again and again we run up against the problem that we cannot formalize the notion of mathematical largeness directly. This is the "problem" which the shadow-casting of the parafinite in its prehistory presents us. How do we capture it? As we will see, the beginning of an answer is supplied by seeing how we *don't*, how the "problem of the parafinite" has fallen through our conceptual nets again and again, only to resurface yet again after a period of historical submersion.

It must be that our conceptual nets are not "right" to capture the problem of the parafinite. Do we have the wrong net, or is it a manner of "throwing away our nets," the way Wittgenstein spoke at the end of the *Tractatus Logico-Philosophicus* of throwing away a ladder after climbing up it?[3] The dominant conceptual net in the foundations of mathematics remains the set theory which finds its inauguration in the work of Georg Cantor. In the third and final framing passage, Alain Badiou invokes this negative relation

---

"*Logic,*" trans. Richard Rojcewicz and André Schuwer (Bloomington: Indiana, 1994), 99.

[2]Gilles Deleuze, *Difference and Repetition*, trans. Paul Patton (New York: Columbia, 1994), 268.

[3]Wittgenstein, Ludwig. *Tractatus-Logico-Philosophicus*, trans. D. F. Pears and B. F. McGuinness (London: Routledge, 1974), 6.54.

between the mathematically large and the conceptual net of set theory:

> *What, however, is an infinite multiciplicity? In a certain sense– and I will reveal why– the question has not yet been entirely dealt with today. Moreover, it is the perfect example of an intrinsically ontological– mathematical– question. There is **no** inframathematical concept of infinity, only vague images of the 'very large'.*[4]

This passage points us directly to several important morals. First, *we must learn to think of the infinite as the long shadow of the parafinite.* First and foremost, Cantorian set theory has no place for the parafinite because it is an *infinite* (or, more appropriately: transfinite) theory. Secondly, and of scarcely less importance, Badiou alerts us that we will find no minimal conceptual formulation of the infinite, and in a sense that I insist remains ineliminably metaphorical, we can only "bound the infinite from below" by "vague images" of the parafinite. 'Vague images' here means something like *the shadow of the parafinite.* But what if our philosophical stake is the parafinite itself? This book does not answer this question directly, but does seek to capture the problem by dissipating the contradictions that have enveloped it.

✳✳✳

This book has been long in coming together. Although its prehistory stretches back to work on Leibniz and the labyrinth of the composition of the continuum in the mid to late 1980's, its specific point of inception was in conversation with Bruce Olberding around 1993 when we were both studying mathematics at Wesleyan

---

[4]Alain Badiou, *Being and Event*, trans. Oliver Feltham (London: Bloomsbury, 2013), 151.

University. I am grateful to the mathematical community at Wesleyan and especially to my dissertation advisor Adam Fieldsteel for their much-needed support. The first writing still included in the final book dates from 1998, and the idea for a volume along these specific lines arose from suggestions made by Angus Fletcher now almost a decade ago. Angus not only encouraged me to write this book but also helped me see *how* to write it, and I owe him especial thanks. I am grateful to all the people, particularly my students over many years, who have contributed to this project.

Portions of this book were first delivered as lectures at Boston University, The University of Georgia, Indiana University, Université de Paris VII, Vassar College, Emory University and West Texas University. I am grateful to the audiences and sponsors for these opportunities. The writing in the Leibniz chapter draws on two previous publications as noted in that chapter, and I am grateful to the publishers for their permissions.

My daughter Zoe Brient and my wife Elizabeth Brient have been a constant source of support, as have the members of my extended family: my brother Timothy Bassler and his family, Marie, Alexander, Ana and Bryan, my brother-in-law Sean Brient, my mother-in-law Mary Brient, and my aunt Ellen Gruenke– all of these among the many, many others. My greatest debt of all is expressed in the dedication of this book to my mother, Shirley Anne Gipson Bassler. Sonam Kachru, Jane Mikkelson and their daughter Alya belong to that family-beyond-family we choose rather than have bestowed upon us.

My first model for work in the history of mathematics and ideas was Salomon Bochner, who served as my mentor during my freshman year at Rice University. My indebtedness to his work and example is primary, and continues. I am also grateful for support from colleagues in the mathematics department at The University of Georgia, and especially to David Edwards and Robert Varley. John Lane Bell's recent Colloquium in the Department of Mathematics at UGA and subsequent conversation provided valuable perspective and inspiration at a late stage in this process, and will accompany me into the next phase of work on these issues. The quantum statistical analogues for the four Principles of the Parafinite at the end of this volume owe much to the thinking, lectures and writings of David Ritz Finkelstein.

Given the wayward history of this prehistory of the parafinite, I am especially happy to have found a home for this project at Docent Press. The support of Scott Guthery and Mary Cronin at Docent Press is most appreciated.

A book written over this many years is indebted to a parafinite number of people, impossible to enumerate here. I thank and apologize to all those I have not included by name; their collective contribution has been integral to this work. All errancy is mine alone.

Athens, Summer 2015

Prologue

Number and Scale

Hans Blumenberg once asked how many readers would make a "sizeable" reception for one of his books. Would 500 do? Or 5,000? Pushing the number higher and higher, he finally declared that if half the people in the world– about 2.5 billion at the time– were to read his book, he would inevitably ask: well, what is the other half doing?[1]

The largest book I know, larger even than the already large books of Hans Blumenberg, is James Joyce's *Finnegans Wake*. Although it's long enough– over 600 pages– it's not nearly the *longest* book I've ever read– Proust's *Remembrance of Things Past* is, cumulatively, considerably longer, and I've heard has been listed in the Guinness Book of World Records as the "longest book." *Finnegans Wake* is certainly also not the most *massive* book I might imagine– one with the entire sequence of the human genome project printed out, say. Recently, Michael Mandiberg has offered a print edition of Wikipedia in 7,600 volumes at a total cost of around $500,000.[2] But Proust and the print version of Wikepedia

---

[1] Hans Blumenberg, "Sättigungsgrade," in *Ein mögliches Selbstverständnis* (Stuttgart: Reclam, 1997), 30.

[2] "Author of the Week," *The Week* July 10, 2015, 22. Although individual volumes are available, the total set is not really available for collective purchase in any very practicable sense. Thanks to Elizabeth Brient for pointing me to

are not, I claim, as *large* as *Finnegans Wake*. A scholar I know
once told me her book was characterized in a review as "path-
breaking," and that her son remarked, "it sure is: you drop it on
the ground and it'll break the path." I like to say there are books
you could kill someone with: literally, not just through boredom,
although there's a large intersection between the two. These are
books which are *massive*. Massive books tend to be long, and long
books tend to be massive, but neither automatically qualifies them
for being *large*. And there are also thin, short books which are
large: Wittgenstein's *Tractatus Logico-Philosophicus* would be an
example.

Clearly I'm trying to get at something which goes beyond simply looking at "large" numbers, where the sense of 'large' actually
corresponds to what I'm calling *long* in the case of books. Part of
the acerbic humor in Blumenberg's little fable is that Blumenberg's
books aren't read by many readers, not so much because they're
long (which they often are) as because they're large. What makes
a book large? Here is a first attempt at a characterization: a large
book is one which is *worked in elaborate detail*. Ironically, what
makes something large is here specified in terms of the small. As
Adorno insists, Wagner's operas are "packed and organized down
to the last sixteenth note" and as such pave the way for Berg's
later work as the "master of the smallest transition" in *Wozzeck*
and *Lulu*, works in which Berg takes no part in his contemporaries'
"opposition to Wagner." Berg's work is no less microscopic than
the shortest works of Webern, but unlike Webern's work its minia-
turization is in the service of large-scale effect.[3]

---

this. She also pointed out the connection to Hans Blumenberg's discussion of
the perfection of the encyclopedia in *The Legitimacy of the Modern Age*, trans.
Robert Wallace (Cambridge: MIT, 1983), 237-38.

[3]Theodor Adorno, *Alban Berg: Master of the smallest link*, trans. Juliane
Brand and Christopher Hailey (Cambridge: Cambridge, 1991), 4. As Adorno
puts it, "Despite a completely different technical approach, Berg is in this
respect related to his friend Webern, whose miniatures are just as intent upon

In a marvelous essay entitled "Pollock's Smallness," the art historian T. J. Clark helps us to see what makes Pollock's paintings feel large. He does this by focusing on some dimensionally small paintings on masonite Pollock did in 1950, which were hung in columns next to much larger canvases in the 1950 exhibition at the Betty Parsons Gallery in New York.[4] Clark draws a distinction between size, which is literal, and scale, which is metaphorical and therefore figurative, and insists that Pollock's paintings are "about" size rather than scale. The paintings "present" size to us by being very finely detailed, and this is more overt, in a way, in the small canvases than it is in the large ones, where we may "get lost" in the sweep.[5] (It's also more overt in abstract painting like Pollock's than it would be in figurative painting, where it could, however, also be true.) This is to say that the large canvases require *more* attention from us, and that we can "practice" by looking attentively at the small paintings. (Practice for listening to Berg by listening to Webern!) Paradoxically: it's *easier* to see that the small canvases are large. Clark begins his essay with a passage from Claude Lévi-Strauss, in which Lévi-Strauss says that a large statue of a human is in fact small because it forces us to see a large-scale rock (or other material) as a smaller-scale human.[6]

When my father used to see paintings by Pollock or paintings "like" Pollock, he would pronounce, as many have undoubtedly done, "I could do that." But in fact he couldn't. I have taken

---

falling silent as the large Bergian forms are upon self-negation" (ibid.).

[4]T. J. Clark, "Pollock's Smallness," in *Jackson Pollock: New Approaches*, ed. Kirk Varnedoe and Pepe Karmel (New York: Museum of Modern Art, 1999), 15-31. I here also acknowledge my indebtedness to this essay as the guiding light in the global revision of *The Pace of Modernity: Reading With Blumenberg*. This revision required a rearrangement of chapters and a rhetorical redistribution which led me to seek the creation of an all-over effect in the presentation.

[5]Clark does not use the language of 'presentation' explicitly, but he does speak about size being "a truly difficult subject," 16.

[6]cited, Clark, 15, from *The Savage Mind*.

to painting post-Pollock abstractions, and I certainly haven't produced any masterpieces but I've learned a lot about the process. One thing I learned is that as a general rule (but no more than that) the more successful canvases will stand up to *close* inspection. My daughter has helped me to photograph some of these canvases and she's declared her domain the photography of what she calls "spots," what would generally be called *close-ups* of details. I've learned a lot about my own paintings (and by implication, about painting) by looking at these photographic close-ups. You can get some sense of the issues involved by trying to find a book with decent reproductions of Pollock's paintings. I haven't found one, really, but the one I find best to look at is, paradoxically, the dimensionally small volume in which T. J. Clark's essay appears. A combination of full canvas and detail photographs does much better than a presentation of full canvas photos alone could ever do. The full canvas photos are consistently underwhelming. I could look into getting some posters, but I doubt this would help much either.

If I tried to tell you what *Finnegans Wake* is about, it would be distinctly underwhelming. (It's about the dark.[7] Proust's masterpiece is about time. *War and Peace* is about Russia.) And if you don't pay close attention, or if you are simply un-sympathetic to Joyce's project, you will find the local detail of his writing irritating and even toilsome. "A linguistic mudpie" is the pronouncement analogous to my father's above. But no, it isn't. If one hasn't the feel for Joyce, or for Pollock, one can simply *miss* the largeness of the projects. And again, in Clark's paradoxical language, Pollock's paintings are large *because* they're small. *Even the dimensionally small-ones.*

Morton Feldman, working in another field—musical composition–and slightly later than Pollock, but sharing many of the

---

[7]John Bishop, *Joyce's Book of the Dark,* **Finnegans Wake** (Madison: Wisconsin, 1986).

same inspirations and dealing with many of the same issues, wrote a four hour piece for small ensemble called "For Philip Guston," in honor of the painter who was his deceased friend. In introducing the piece to a group of people who were about to sit through a performance of this mammoth, he had this to say, in conclusion:

> [*Aside*] It's a short four hours. I now am paying the price. I have a Greek eternal punishment. I have to sit through these things [*all erupt into laughter*]. And there are some– there is an hour-and-ten-minute piece which is a *very long* hour-and-ten-minute piece– but this? This piece doesn't give you the feeling that it's four hours.
>
> OK. Thanks.[8]

Is Feldman saying about his piece the opposite of what Clark is saying about Pollock, or is he saying the same thing, or something else altogether? Getting such issues into focus is part of the difficulty we face here: our language is inadequate. Kant tried to approach problems historically antecedent to these, but analogous, by distinguishing between the beautiful and the sublime, calling the sublime the "absolutely large." The topic of the sublime had been a hot spot for culture since the appearance of Edmund Burke's *Philosophical Inqiiry into the Origin of Our Ideas of the Sublime and Beautiful* in 1757. Yet to do justice to the experience of size we feel in viewing Pollock's work or listening to Feldman's, we need to move beyond the vocabulary of the sublime, which is representative of a certain stage in the development of modern aesthetic culture. There's something much closer to our contemporary experience of largeness, of what I will come to call the parafinite, in this work closer to our contemporary condition.

---

[8]Morton Feldman, *Give My Regards To Eight Street: Collected Writings of Morton Feldman*, ed. B. H. Friedman, afterword by Frank O'Hara (Cambridge: Exact Change, 2000), 200.

So what distinguishes Pollock from bad imitations—including my own?! No, it isn't just Pollock's reputation. Clark, who has been looking at Pollock for decades, finds some of Pollock's canvases distinctly more successful than others, just as we would with any figurative painter, and I happen to trust him. (Not that we need agree with him about which these are; Kant rightly insisted that in the aesthetic domain we mustn't ever expect such ultimate agreement, even if we can take it as a kind of ideal. And, more particularly, what feels large to me may not feel large to you.) But this is a question looking for a quick fix, because the question is what makes Pollock's paintings good, and this is an instance of the question what makes a painting good, and we shouldn't expect any simple answer to this question. There's a wonderful story, though, which Stan Brakhage told. Pollock painted in a large barn, and when someone spoke abusively of his paintings, despite his (apparent) drunkenness, Pollock pointed to a doorknob at the opposite end of the barn and then flicked a jet of paint across the length of the barn and hit the doorknob spot-on.[9] This sounds like mythology, but mythologically it gives us an answer– which is no literal answer– to our question.

Ultimately, I intend the parafinite to be a notion of largeness in something like the sense in which T. J. Clark wants to talk about *size* rather than *scale*. Pollock's painting became representative for an entire epoch of American history (and, like Clark, I am happy to join the "rogues' gallery" of historicists) because somehow it captured the portentous changes which in some sense made the 1950's *the* American decade (and goodbye to all that). As I intend it, the notion of the parafinite is something we are headed *toward*,

---

[9]The story is told in special features included in the Criterion Collection edition of Brakhage films: *By Brakhage: An Anthology*, 2 vols., CC1590D and CC1898D (The Criterion Collection, 2003, 2010). Brakhage's painted films constitute a privileged locus for extending considerations of size and scale in Pollock's paintings.

something somehow in our conceptual "blind spot,"[10] and the "long shadow of the parafinite" is the shadow which the notion of the parafinite casts *back* upon us from the future it inhabits, somewhere out ahead of us on the road of modernity. This idea, too, is not a new one: it comes straight out of medieval Biblical exegesis, in which coming events from the New Testament "cast their shadow before"[11] upon the events of the Old Testament, and the Second Coming casts its long shadow on our current condition. (This is why Kierkegaard insists that his notion of repetition is "repetition forward" rather than backward, which would be recollection rather than repetition. Repetition makes *past* events be what they are *for the first time.*[12])

From our modern vantage, we tend to think of human history as constantly accelerating, continuously becoming larger and faster.[13] But in fact, human history is becoming *discontinuously* larger and faster, and we are in a very good historical position to see just the nature of this discontinuity. In his powerful book, *Ecological Im-*

---

[10] Jean-Yves Girard, *The Blind Spot: Lectures on Logic* (Zürich: European Mathematical Society, 2011).

[11] The phrase is stock; see e.g., James Joyce, *Ulysses* (New York: Vintage, 1986), 135, and also Richard Barber, *The Holy Grail: Imagination and Belief* (Cambridge: Harvard, 2004), 152, where Barber enlists the phrase to support the idea of Grail narratives as imaginative reworkings of scripturally based traditions.

[12] And, even more paradoxically, present events, too. The notion of "conditioning on the future" is perhaps exotic, but not at all difficult to handle, in contemporary physics. For a quantum-theoretic explanation of the "arrow(s) of time" in terms of future conditioning, see L. S. Schulman, *Time's arrows and quantum measurement* (Cambridge: Cambridge, 1997); Schulman's application of future-conditioning is inspired by work of Thomas Gold, in *The Nature of Time* (Ithaca: Cornell, 1967).

[13] For a view in some regards contrary, see George Steiner, *In Bluebeard's Castle: Some Notes Towards a Redefinition of Culture* (New Haven: Yale, 1971), esp. Chapter One, "The Great Ennui," 3-25, in which Steiner discusses a nineteenth centure sense of the collapse of the quickening of time felt prior to 1815. Steiner's caveat is specific to the nineteenth century.

perialism: The Biological Expansion of Europe, Alfred W. Crosby forcefully argues that the "Neolithic Revolution" ending approximately 3000 BCE concludes an era of dramatic expansion that is unequalled by anything in the next 4000 years. "Nothing in those four millenia compares in importance with the domestication of the horse, for instance," and already by 3000 BCE humanity had "produced its greatest practical botanists and zoologists of all time."[14] From a less biological and more intellectual perspective we might insist that all sorts of novel things happen in the 4000 years Crosby sees as relatively static. But in fact, there is no real tension here, for at almost every level of historical scale, periods of intellectual development are often, though certainly not always, coterminous with periods of practical consolidation.

A new era of revolution begins between 900 and 1900 CE Crosby hesitates to assign a precise "date of birth" to this new modern revolution, but "[i]n the year 1000 A. D. (or at least during the century or so in which that year falls), Western Europe ceased being the wrack left behind by the ebb of the Roman Empire and began being something new and vital," marking "the first birth of a society of remarkable energy, brilliance, and arrogance. Such societies are often expansionistic."[15] Crosby's portrait is a sobering one in which the features of modernity on which we have come to rely are integrally connected to an expansionism which duly qualifies for the name 'imperialism'. However we may attempt to cope with this acceleration, I think Crosby successfully argues that we are dealing with something of recent vintage which is, at the same time, not unprecedented in human history, nor is it simply a constant of the human condition.

---

[14] Alfred W. Crosby, *Ecological Imperialism: The Biological Expansion of Europe, 900-1900* (Cambridge: Cambridge, 1986), 42 and 20 respectively. On traditional zoology and botany, see also Claude Lévi-Strauss, *The Savage Mind* (Chicago: Chicago, 1986), 38-46.

[15] idem, 44.

For the most part, these sorts of concerns[16] for the parafinite are well beyond the bounds of this volume, however, which will focus on the more tractable issue of large numbers as a central, simplifying example. In the case of numbers, the *really* large ones are always on the other side of those we are currently counting. This is striking because it points out that a natural notion of symmetry breaking is embedded in our conception of the numbers. There are always finitely many numbers before any number we choose, and "infinitely many" (whatever that means) that come after. Further (and excusing the pun), as numbers grow larger and larger, they become in some sense harder and harder to count, as is made evident by the fact that we need to find newer and more powerful ways to write large numbers down. Decimal notation was a major innovation: it made writing 1,000,000,000 almost as easy as writing 10. But by contemporary mathematical standards, a billion is a very *small* number. "Really" large numbers are much harder to count: they are the wild, Wild West of counting, and throughout our history we have needed to devise more and more powerful ways to name larger and larger numbers. Now, as computing technology proliferates and world population reaches an unprecedented size, the demand to "handle" large numbers becomes greater than ever. And yet the traditional formulation of mathematics largely obscures the foundational significance of large numbers and the problems associated with counting them. From a perspective which attempts to provide a foundation for mathematics, all the counting numbers, large or small, are on a par with each other. It makes no difference whether we are talking, say, about 13 or 13 billion, or even something "much" larger. Consequently, there has been no philosophical answer to the question: what makes a number large?

If we can understand why it has been so difficult to "see" the

---

[16]I have attempted to provide the first rudiments of a vocabulary to deal with them in *The Pace of Modernity: Reading With Blumenberg* (Melbourne: re.press, 2012).

problem of giving a theoretical description of the special role of large numbers, then we will have taken one step toward resolving the divide between how mathematics is understood and how it is used. Traditionally, the largeness of numbers has always been understood as something relative to the individual practical contexts of their application: seven is a large number of labors of Hercules, but usually not a large number of beans (unless it's the number of beans you've put in your ear). Typically, mathematicians would say that whether seven is large is not a mathematical issue at all, but *only* an issue of application, and this attitude has a great deal of plausibility. But it leaves us without an account of largeness per se. Indeed, a foundational mathematical account of largeness would force a rethinking of this very distinction between "mathematics" and "application." In the mathematical domain, I call the subdomain of large numbers *the parafinite* and the problem of accounting for large numbers *the problem of the parafinite*. In this book, I try to show how the problem of the parafinite casts a long historical shadow over mathematics and our understanding of it: it is only "seen" indirectly, by the shadow it casts. The limits of my mathematical language are the limits of my capacity to count: how are we to *write down* names for large numbers? This question about mathematical language, in particular, brings the problem of largeness closer to direct recognition. But what "looks" in a given language like a large number is really only a shadow-casting, a particular expression of largeness in a particular number, and not the idea of largeness itself. With an improved language, any such number will come to seem quite small. The way a given number divides the counting numbers up into those smaller and those larger than it is really only like the boundary of a shadow, which separates light from dark. We cannot say that any particular number *is* large, for there are always numbers "much" larger. The large numbers are always the ones "which follow," which fill in the dots after 1, 2, and 3. So how are we to capture linguistically the essential feature of numbers, that they provide answers to the question, "How large?" There is something peculiarly indirect about our capacity to speak of the largeness of numbers, and we can take the first step in battling this indirectness by adding to our language a name for the domain of large numbers: the parafinite.

Imagination is always ultimately outstripped by large numbers, but we need to cultivate an indirect imagining of the parafinite: we need to learn to spot its shadow in the limits of our mathematical language. In the aftermath of the twentieth-century crisis in the foundations of mathematics (and beyond), we are now, once again, on the cusp of an explicit, and powerful, recognition of the problem. Contemporary mathematicians do not work in relative isolation, the way Archimedes had to in an ancient context. Nor do they face the theological pressures constraining early modern practitioners like Galileo and Leibniz, which forced them to couch their radical ideas in ways which softened their impact. Now, we face a different, but no less challenging, set of problems: new ideas are cast adrift in a sea of information glut, and innovations have to be "discovered" over and over again. Market pressures dictate that new ideas must compete at the auction block, and are guilty until proven saleable. These issues of information explosion and economic competition themselves require a language of largeness and scale for us even to begin to express them adequately.

New ways to talk about the parafinite in a modern context have been glimpsed by various visionaries, but we now require a systematic development which must go beyond these radical manifestos: what is required is a fully developed language of the parafinite. The parafinite and the problems which surround it need to be written into the foundations of our various cultures, intellectual and otherwise. For we live in a world of ever larger numbers, from the concrete anonymity of mass-marketing statistics to the ultra-theoretical reaches of prime number encryption. The gigantism of our technological culture has disconnected it from its scientific roots. Historically, this disconnect results when the language of science is outstripped by the proliferation of technology. We need to update the language of science to deal with these demands, and that updating must begin with a retooling of the language of large numbers. Throughout the rise of science and technology this has been an ongoing need, which has been tackled in ad hoc ways for localized areas of concern. But these patch-up jobs will only work for so long. What we need now is an overhaul from the ground up. We need a language which can handle the concept of the parafinite. And we can begin this venture by looking into childhood education.

In 1957, a Dutch schoolteacher named Kees Boeke published a children's book, *Cosmic View: The Universe in Forty Jumps*, which toured our cosmos over forty tenfold shifts of scale, from the largest reaches of the cosmos available at the time down to the smallest microscopic depths.[17] This little book came to the attention of the renowned husband and wife designing team of Charles and Ray Eames, and about a decade after it was first published they had refashioned Boeke's jumps into *A Rough Sketch for a Proposed Film Dealing with Powers of Ten* (1968), which another decade later would spawn the 1977 short film, *Powers of Ten*. Five years later the theme would jump back into book format when Philip and Phylis Morrison collaborated with the Eames' office on their own book, *Powers of Ten*.[18] Over a quarter of a century,

---

[17]I thank my former student Gordon Lamb for introducing me to the Macdonald First Library book, *Number*, which ends with a discussion of scale in terms of "place value," (London: Macdonald Educational, 1970). I keep it next to one of my childhood favorites, *The Story of Flight* by Mary Lee Settle, illustrated by George Evans (New York: Random House Step-up Books, 1967), which has a chapter called "Farther and Faster."

[18]Philip Morrison and Phylis Morrison and the office of Charles and Ray Eames, *Powers of Ten: About the Relative Size of the Universe* (New York: Scientific American Books, 1982), vi-vii. The two Eames films are available on Volume I of *The Films of Charles and Ray Eames* ID921OPIDVD (Lucia Eames DBA Eames Office, 1989), distributed exclusively by Image Entertainment.

the idea of presenting the cosmos according to its scales jumped from book to film and back again, each time gaining in elaboration. But the basic idea is simple: to present us with a sense of awe at the many scales over which we have come to know the cosmos we inhabit.

There is nothing new about this sense of awe in itself. Aristotle famously (for a philosopher, that is) declared that philosophy begins in wonder, and our wonder at the world we inhabit is at the heart of this sense. Ancient cultures (and India in particular) were piqued by the game of naming ever larger numbers; very early in Indian cultural history learned pundits knew a method for estimating the number of leaves on a tree starting from the number of branches, and they developed systems for representing large numbers long before the Greeks. The ancient Chinese referred to the "ten thousand things" as a general term for an expansive multitude, and the Greek mathematician Archimedes followed out the question of naming the number of grains of sand in a universe of our size (as he estimated it). The outcome of Archimedes' exercise, obvious in its appearance but not in its implications, was the capacity to name numbers even larger than anything which would express physical sizes in Archimedes' universe.

The tradition of naming large quantities (sizes) and magnitudes (distances, for example) is carried on in the early modern period as a concomitant of the interest in the expanse of the cosmos. In the seventeenth century, Robert Burton, in his *Anatomy of Melancholy*, reaches beyond the classification of human psychology into a generalized taxonomy which leaves few stones of then contemporary interest unturned. So it should come as no surprise that among his digressions we find one on the capacity to calculate large quantities, mentioning three problems canonical for the early modern period. The first he discusses is Archimedes with his grains of sand. The second problem is to calculate how many people could inhabit the earth: "some say 148,456,800,000,000,"[19] then multiplying this by the number of generations spanning 60,000 years, giving 89,074,080,000,000,000,000, a sizeable enough number. The third problem Burton mentions leads to much greater magnitudes still, ones which would concern Leibniz a little bit later: how many words may be formed out of the letters of the alphabet, and how many books could be formed out of these words?[20] This "variation" leads to quantities which Burton insists "will not be contained within the compass of the firmament."[21] Key to all these examples, in one way or another, is the phenomenon of exponential rather than additive growth. That is, instead of growing by adding (2, 4, 6, ...) or growth by multiplying (2, 4, 8, ...), we grow by the "next" available operation, exponentiation, which consists of repeated multiplication just as multiplication consists of repeated addition: 2, 4, 16, 256, 65536, ... For all sorts of reasons, exponential growth, as it turns out, is very special indeed. It marks a kind of barrier reef for the dynamics of divergence, beyond which we move into the waters of the mathematical deep.

The capacity to describe, and hence in the broadest sense to name, large numbers seems to be a prerequisite for imagining them (but perhaps not quite for thinking about them). A very powerful technique, which uses the power of exponentiation in a limited way, is provided by the aforementioned decimal system, which was not yet fully developed by the ancient Greeks. Archimedes' attempt to count the number of grains of sand needed to fill the universe resulted in something like it. But the techniques for naming large numbers quickly outstrip our capacity to imagine them. In a way,

[19]Robert Burton, *The Anatomy of Melancholy*, ed. Floyd Dell and Paul Jordan-Smith (New York: Tudor Publishing Company, 1927), 460.

[20]Gottfried Wilhelm Leibniz, *De l'Horizon de la Doctrine Humaine:* Ἀποκατάστασις πάντων *(La Restitution universelle) (1715)*, trans. and annotated Michel Fichant (Paris: Vrin, 1991). This volume documents the prehistory of Leibniz's preoccupations as well.

[21]Burton, *Anatomy,* 460.

this is not surprising if you think about the "bookkeeping" function of numbers: the whole point of adding up a list of numbers is that then you don't have to carry them all around in your head. But as the technical innovations for naming numbers progressed, an increasing human alienation from the realm of mathematical quantity can also be felt, and the recent efforts of Boeke, the Eames and the Morrisons, like the earlier efforts of Archimedes, Burton and Leibniz may be seen as cultural attempts to address not only the wonder we feel at the massiveness of the universe but also the alienation we ultimately come to feel in the face of a mathematics so powerful it outstrips our own capacity to imagine.

Behind the initiatives of Boeke, the Eames and the Morrisons lies the psychological trick of what we may call the *visual leap*. It is greatly akin to something we all know from film, the technique of the "jump cut," but it adds to it a transition to a different scale of largeness or smallness. Well before *Powers of Ten*, we can find examples of it in literature. In his utopian cosmology, *Eureka*, Edgar Allan Poe tricks his reader in a way which dramatizes the technique of the visual leap. After declaring that in order to conceive of the distances which separate our solar system from "one of the brighter stars" we would already need "the tongue of an archangel,"[22] Poe nonetheless goes on to attempt to provide some feeling for such distances through the use of relative proportions. Suppose that instead of being 95 million miles from the Sun, as Poe tells us it is, our Earth were instead only 1 foot away from the Sun. If the relative distances of the planets were preserved, Neptune would still be 40 feet distant, *"and the star Alpha Lyræ, at the very least, 159."*[23] Poe doubts that the reader will have found anything objectionable in this last claim.

But my account of the matter should, in reality, have

---

[22]Edgar Allan Poe, "Eureka," in *Poetry, Tales, & Selected Essays* (New York: Library of America College Editions, 1996), 1335.
[23]idem, 1336.

run thus: – The distance of the Earth from the Sun being taken at one foot, the distance of Neptune would be 40 feet, and that of Alpha Lyræ, 159– *miles*: – that is to say, I had assigned to Alpha Lyræ, in my first statement of the case, only the 5280th *part* of that distance which is the *least distance possible* at which it can actually lie.[24]

Poe has attempted to give us some sense of the dramatic distance by playing on our sense of scale, and in particular the *jump in scale* which is needed to imagine the distance to the stars *even relative to the largest distances in our solar system.* In order to imagine these distances, we are forced to shift scale, and the trick Poe has played on his reader drives this home. In fact, Poe still dramatically underestimates the distances involved: it was not until well into the late 19th and early 20th century that these measurements began to stabilize.[25]

Throughout the modern period there is a close historical connection between the concern with the cosmos and the concern with imagining large quantities: the imaginative concern with the large seems to precede the concern with the microscopically small. On the mathematical as opposed to the cosmological front, the situation differed in two significant regards. First, mathematicians were much more concerned with the infinitely large and small than they were with the finitely large and small. Throughout the modern period it has been more or less "mathematical orthodoxy" that numbers are not large or small "in themselves" but only relative

---

[24]idem, 1337.

[25]Alpha Lyrae, also known as Vega, is currently measured at roughly 25 light years from earth. Since the sun is roughly eight light minutes from earth, using these estimates and taking the distance from earth to sun to be one foot we get 311 miles for the distance to Vega. The calculation is very rough as I've rounded everything to whole numbers. Poe's estimate is about half of this current one.

to each other. 5 is smaller than 7 but larger than 3. 26,000 is larger than all of these but smaller than 26,001 or 534,678. Consequently, when issues of "largeness" and "smallness" were debated by mathematicians, the focus was on the "infinitely large" and the "infinitely small." By way of contrast, physical magnitudes could at least be recognized as "large" or "small" in kind: we can say what constitutes a tall person (name your favorite basketball giant) or a large planet or a large star. These are empirical matters, and since they are subject to change, it's not so immediately clear what distinguishes this sort of situation from the utterly relative largeness and smallness we find in the mathematical domain. But it's a historical fact not much worth debating that when people have thought about largeness and smallness they've tended to focus on finite sizes in the world and infinite and infinitely small sizes in mathematics.

Furthermore, for complicated and still somewhat murky reasons, the infinitely large and the infinitely small were historically not "on a par" so far as these debates were concerned. In mathematics it was the status of infinitely small magnitudes, so-called "infinitesimal" line segments, in particular, which came to dominate debates about the status of the mathematical distinction between the finite and the infinite in the early modern period. Informally, an infinitesimal quantity is one that has some length but is smaller than any finite length. Exactly what this is supposed to amount to, however, generated a great deal of controversy. In the eighteenth century Berkeley caustically referred to Newton's infinitesimal "fluxions" as the "ghosts of departed quantities," and he was one of the early driving forces in a witch hunt to eliminate infinitesimals from the mathematical domain. The debate raged on, literally for centuries, until late in the nineteenth century Georg Cantor's set theory banished these informal infinitesimal magnitudes more or less definitively. However, Cantor's elimination of infinitesimals came at a high cost. In order to banish infinitesimals he had to commit to a strong conception of the infinite. Ultimately, this conception of the infinite presented almost as many philosophical, if not mathematical, problems as the deposed infinitesimals. To this very day, just what this concept of the infinite amounts to continues to be a subject of philosophical debate.[26] On the face of it, then, it's not so clear what is gained, really, by eliminating the infinitesimal if it requires us to *espouse* the infinite. Why be so concerned with eliminating the infinitely small if we do so by appealing to the infinitely large?

The answer to the question begins with a distinction I have already mentioned above, between *quantity* and *magnitude*. A quantity is a number of things; it tells the size of a collection. Thus the simplest quantities are numbers like 1, 2 and 3, and we say things like "there are three apples on the table." (Incidentally, the Greeks didn't admit 1 as a quantity since they thought only multiplicities had a quantity, and to have a multiplicity you need at least two things. Also, they said something which was closer, e.g., to "there is a three of apples on the table.") Fractions can be thought of as quantities, too, if we think in relative terms: $\frac{1}{2}$, for example, tells the ratio of one quantity to another twice bigger. If there are four apples on the table, two of them make up half of the apples. What about magnitude? A magnitude, in contrast to a quantity, is something with a particular "dimensionality," like a line, say, or an area or volume. We say that a length is a one dimensional magnitude since it stretches out in one dimension. An area, on the other hand,

---

[26]Shaughan Lavine, *Understanding the Infinite* (Cambridge: Harvard, 1994). Also, since Abraham Robinson's work in non-standard analysis, infinitesimals are back again on a "firm," Cantorian footing. These are, however, *formal* infinitesimals, and as such are categorically distinct from the pre-Cantorian infinitesimals. See Abraham Robinson, *Non Standard Analysis* (Amsterdam: North Holland, 1966), repr. Princeton 1996. Nowadays there are many different "post-Cantorian" formal approaches to mathematical infinitesimals. Informal approaches continue to be used by many scientists and engineers (and in their less guarded moments, even by mathematicians).

would be a two dimensional magnitude, and a volume would be a three dimensional magnitude. Although our physical world only has three spatial dimensions, we can still think of higher dimensional magnitudes as well (like a four dimensional "hypervolume" in spacetime, for example).

In fact, Cantor's concern with infinitesimals wasn't really that they were infinitely *small* at all. His concern was that they were *magnitudes*: when mathematicians spoke of infinitesimals, they would have in mind infinitely small lines, or areas, or volumes. Infinitesimals "stretch out." Quantities, on the other hand, if you think about it in the strictest sense, don't even exist in space and time. Although four *apples* exist in space and time, right here on the kitchen table, say, "four" is a concept, and concepts don't live in space and time at all. Magnitudes, on the other hand, *aren't* concepts, or at least they aren't *obviously* concepts. We might *try* to make sense out of the idea of a concept of "stretching out," but Cantor for one wasn't at all sure what that could mean, and he had a point. To put it a little more precisely: the concept of four *units*, i.e. a collection of four, was something that Cantor felt comfortable with (as it turns out, philosophers have found problems here, too–as they most always do). The concept of something with length four causes more trouble. To begin with, you have to *specify* a unit in order to talk about a length: something isn't just "four" long, but rather four inches or four feet or four ells or four kilometers. But if you have to talk about units *anyway*, then eliminating magnitudes offers an obvious advantage, since if you eliminate them you *only* have to talk about collections of units.

So Cantor didn't have any problem with the infinite per se, whether in the small or the large. What he had a problem with were magnitudes. This, too, is nothing new: even the ancient Greeks, who worked largely with geometric magnitudes, recognized up to a point the need to distinguish quantity and magnitude. But in the early modern period, in particular, the birth of calculus was ushered in through a persistent playing fast and loose with magnitudes and, in particular, infinitely small ones. This was Cantor's real target, since his goal was to put the foundations of mathematics on a "rigorous" footing. The fact that he ultimately lacked a "rigorous" justification for his appeal to the infinite did not seem to bother him. Indeed, his justification for the concept of the infinite was largely in theological terms that, for better or worse, even specialists in the foundations of mathematics usually pay little attention to today. Instead, it is embodied in the so-called "Axiom of Infinity," which remains a subject of philosophical debate.[27]

However, all of this does make matters tricky for understanding the historical development of attitudes toward the infinite. Historically, debates about the distinction between the finite and the infinite are tangled up with the distinction between quantities and magnitudes. For now, let's simplify matters by focusing exclusively on quantity and consider the question: how do we count large numbers? Here, the operation which plays something like the role of the visual jump shift described above can be described in terms of iteration, or self-application. Without our thinking much about it, this process is already embedded in the decimal system we use to represent numbers all the time. But to see this, let's start from a more primitive point.

Perhaps the "simplest" way to represent numbers is in terms of what is often called "stroke notation." This just means that a number of strokes is used as a "name" to represent a given number (notice how the word 'number' appears twice in this sentence before the parenthesis). In stroke notation, the connection between the notation and the number itself is so close that it borders on identity. The first several stroke numerals, i.e. the "names" of the first several numbers, are given as follows:

$$|, ||, |||, ||||, |||||, ||||||, |||||||,$$

---

[27]Here, again, see Shaughan Lavine, *Understanding the Infinite*.

where the commas in the above line separate the different stroke numerals representing the numbers between one and seven inclusive. Even though this notation "exemplifies" the concept of number, since each collection "has" the number of strokes which it represents, here we will still insist on distinguishing between the "numeral" and the "number" which it names. When we use more sophisticated notation systems, it will actually be easier to see this distinction. For example, the *numeral* '5' involves one symbol while the numeral '27' involves two, whereas we can roughly say that the *number* 5 involves five "things" and the *number* 27 involves twenty-seven of them. (Notice how when I'm talking *about* the numeral I put the symbol in single quotes, thus: '5', whereas when I'm *using* it to refer to the number I write it straight out, without the single quotes.)

In the case of the stroke notation system, the name for the number one has one stroke, the name for the number two has two strokes, etc. Note, however, that in this notation system not just any collection of four things will do to represent the number four: four blotches or, say, four triangles. In this sense we can see that the stroke notation system amounts to more than simply collecting together any four things to represent the number four.

As you can see (almost literally!), it will soon become very cumbersome to represent numbers using the stroke notation. The first trick we can use to make things easier is to group strokes. For example,

|||||||||||||||||||||||||||||||||||||||||||||||

is difficult to take in "at a glance," and even if you try to count it carefully you might make a mistake. It's much easier to tell that

||||| ||||| ||||| ||||| ||||| ||||| ||||| ||||| ||||| ||

amounts to forty-seven strokes. To see this, however, we count nine bundles of five strokes, multiply nine times five, and then add two. This already involves not just the operation of counting along one by one, which we will call the "successor" operation: three is the successor of two, four is the successor of three, etc. It also involves the operations of addition and multiplication. Of course, you *could* just count straight through one unit at a time, but then you would lose the advantage of bundling.

The use of multiples is integral to the Roman numeral representation of numbers, where forty-seven would be given as

XXXXVII.

Here each '**X**' stands for a bundle of ten in the stroke notation, and '**V**' stands for a bundle of five; the two symbols '**II**' function like two strokes. Thinking of it this way is a way of "converting" the Roman numeral notation "into" the stroke notation.

Yet moderately large numbers are already cumbersome in Roman numeral notation, and for larger numbers our decimal system works much better. The decimal notation system is a *positional* notation system, which means that the values of the symbols depend on their respective positions. For example, the decimal numerals 423 and 324 have the same components, but they represent different numbers because we find the symbols '2', '3' and '4' in different positions. To understand the value of these symbols fully, we must have recourse to the notion of exponentiation along with addition and multiplication. For example, '423' means "four times 100, plus two times 10, plus three," whereas '324' means "three times 100, plus two times ten, plus four." With each increase in position we *jump shift* by a *power* of ten. In particular, $100 = 10 \times 10$, which is "ten squared," whereas $10 = 10 \times 1$. Each time we move to a higher power in the decimal system we use a higher power of ten. These powers of ten are the outputs of exponentiating the number ten:

$$10^1 = 10; \ 10^2 = 100; \ 10^3 = 1,000; \text{etc.}$$

Thus our decimal representation involves the operations of successor (adding by one), addition, multiplication and exponentiation.

As it turns out, there is a way to understand each of these operations as an iteration, or "self-operation" on the preceding one. First, consider the operation "successor," which involves adding by one. Suppose we start with some number, say 5, and iterate, i.e. "perform," the successor operation three times: "6 is the successor of 5," "7 is the successor of 6," "8 is the successor of 7." By repeating the successor operation three times, we have effectively added 3 to 5, obtaining 8. That is: addition is iterated, or repeated, succession.

Similarly, multiplication is repeated addition. Suppose we start with zero and add five three times: "0 plus 5 is 5," "5 plus 5 is 10," "10 plus 5 is 15." In this way we achieve $3 \times 5 = 15$ as a process of repeated addition. In turn, we may see exponentiation as repeated multiplication: the square of 10 is two factors of 10 multiplied together, the cube of ten is three factors of 10 multiplied together, and in general, 10^n is n factors of 10 multiplied together, where n is some counting number, called in this case the "exponent of 10."

What about if we iterate the exponential operation? This already leads to a much less familiar operation, usually called "hyperexponentiation." For this operation, we already need a new notation. The most common one uses a double upward arrow. For example, let $2{\uparrow}{\uparrow}5$ consist of five 2's in an exponential "stack." That is,

$$2 \uparrow\uparrow 5 = 2^{2^{2^{2^2}}}.$$

Let's figure out what this number is.[28] Start at the top of the exponential stack, with the two topmost 2's. Since $2^2 = 2 \times 2 = 4$, we have that

$$2 \uparrow\uparrow 5 = 2^{2^{2^{2^2}}} = 2^{2^{2^4}}.$$

---

[28]Strictly speaking, the notation as written is ambiguous, because in general, $a^{(b^c)} \neq \left(a^b\right)^c$. For stacks without parentheses written explicitly, I will adopt the convention that we should read down from the top of the stack. Thus, e.g., $a^{b^c} = a^{(b^c)}$.

Now $2^4 = 2 \times 2 \times 2 \times 2 = 16$, so we have in turn that $2 \uparrow\uparrow 5 = 2^{2^{16}}$. 2^{16} is 2 multiplied by itself 16 times, and this turns out to be equal to 65536, so we have, finally, that $2 \uparrow\uparrow 5 = 2^{65536}$, which is 2 multiplied by itself 65,536 times! Don't try this at home: just writing out such a number in decimal notation would require more paper than you could fit into what we've observed of the universe so far! Indeed, it's (much) larger than the estimated number of fundamental particles in the observable universe (we don't know what's beyond what we've seen, or indeed whether the universe is finite or infinite, though there are some grounds for debating this question.) That means that this number is (much) larger than any meaningful *physical* quantity.

This number is indeed so large that one mathematician, Edward Nelson, has declared that it is actually infinite! Not many people agree with him about that, but it does drive home the point that we're talking about a *really big* number– at least in some physical sense. And yet most mathematicians would have it that there's nothing "intrinsically" large about this number *at all*. After all, it's smaller than $2^{65536} + 1$!

Even if this number is only large relative to the finite numbers that we usually encounter in the physical and mathematical "worlds," it is striking that we could describe this number using very small components, namely 2 and 5, and only four iterations of that most basic of operations, the successor operation! This demonstrates just how powerful a tool iteration is.

Just as we achieved hyperexponentiation from an iteration of exponentiation, so we can achieve hyperhyperexponentation by iterating hyperexponentiation, and so on and so forth: there is no end to the process– unless there were to be an end to the counting numbers! We need a new notation to organize this ongoing process of iteration. To do this, we will employ the now familiar device of *re-symbolization*, just as we did when we went from the stroke numerals to the Roman numerals, or to the decimal numerals. Each time we do this we are effecting a *notational jump shift*.

Call

$$g_1(x, y) = x^y = x \uparrow y;$$
$$g_2(x, y) = x \uparrow\uparrow y;$$
$$g_3(x, y) = x \uparrow\uparrow\uparrow y;$$

and so on and so forth, where each function is the iteration of the previous one. (As I generally say to my students right around this point: are we having fun yet?) As we've already seen, $g_2(2, 5) = 2 \uparrow\uparrow 5 = 2^{65536}$ is already a *huge* number, at least physically speaking. And each of these functions spits out numbers which are screamingly larger than the previous function does. You might therefore ask: are there any functions that spit out even larger numbers than all the functions in this list? The answer is essentially yes, but since the process is somewhat involved I'll defer it to Appendix A.

Even so, you may also want to ask another question: aside from a feeling of mathematical dizziness, what does all of this get for us? Or, to put the point a little more specifically: if 2^{65536} is already so far beyond our capacities to imagine numbers, what's the point of generating numbers much larger still?

Good question.

In the rest of this introduction, I will supply a series of preliminary answers to this question, and these answers will serve two purposes. First, they will mark out the exoskeleton of the rest of this book. Second, they will provide an indication of a competition between the notion of the large but (traditionally) finite and the notion of the (traditionally) infinite. This battle sets the stage for my main concept in this book: the concept of the parafinite. To call it a concept may not be quite accurate: it's more like a suggestive name for the lack of a definite concept, a "blind spot" in our mathematical orientation. But we'll see soon enough what we're in for.

My first answer to the question why large numbers matter has to do with practical ways (perhaps *all* practical ways) in which we use mathematics. When I was younger, and calculators were still novel and exciting, you used to see some people at the grocery store tabbing up their purchases on a calculator as they strolled through the isles dumping groceries in their carts. I haven't seen that in quite a while, but one way or another most everybody goes through the checkout lanes. I prefer U-SCAN myself, but either way somebody (one might say some*thing*, but even in the U-SCAN case *I* am still involved) adds up the prices of all my groceries and I pay a sum of money equal to the number they (/it) arrive(s) at.

Now, you wouldn't have to do it that way. That is, although it would be terribly impractical, you could pay for the first item, then pay for the second item, and so on (for current purposes, let's forget about tax). Why don't we do that? Of course, because it would be *impractical*. It's *practical* to add the prices together and pay for all the items at once.

Suppose you have twenty items in your cart (in that case, please do NOT go through the express lane). In some larger mathematical sense, twenty is not a particularly large number. But it would be an annoyingly large number of individual transactions. So for practical purposes, twenty is large enough that it's better to add the prices of that many items together and pay for all of them at once.

The grocery store scenario illustrates a principle that was already at work in our discussion of notation systems above. In that context, it quickly became "annoying" to count out long strings of strokes to identify what number was being represented, and so it was more "convenient" to group them into groups of five, since each group of five can be taken in "at a glance." (This is not *altogether* different from how, if you're not up for it, it's annoying to try to read *Finnegans Wake*.) But then to manipulate these we are already involved with the operations of addition and multiplication. In this case of stroke notation, it is even clearer how our grouping, adding and multiplying are a function of limits on our capacity to take collections of strokes in "all at once," or "in a glance." Even "Rain Man," who could immediately tell you how many matches fell out of a matchbox, would have limits in this regard, though they would be somewhat higher than normal. Both examples make a point that the mathematically trained, but psychologically attuned philosopher Edmund Husserl was at pains to emphasize early on in his career: if there were no limits to our capacity to hold numbers directly in our heads, there would be no need for mathematics. Husserl put the point somewhat differently, and more dramatically, by saying that God would have no need for arithmetic.[29] But either way, the point is simple and powerful. Even so, or maybe even because it is so simple, it is an easy point to forget (philosophers do, all the time).

Perhaps the main reason philosophers are especially prone to ignore this point is because they almost always view it as a "merely practical" or "merely psychological" fact. This makes it sound like the point has no "theoretical" bearing, no implications for a philosophical account of the nature of mathematics. By doing this, they effectively vote such practical matters, based on psychological limitations, out of the philosophical arena. If they do happen to think that such practical matters and psychological facts are important, they usually do so because they consider themselves "pragmatists" or "naturalists": philosophers for whom the "cash value" of empirical givens plays an ineliminable, perhaps even foundational, role. But even these philosophers, like their more traditional counterparts, would tend to deny that there is anything intrinsic to the nature of numbers themselves (as opposed, say, to their application in "practical" contexts) which embodies the structure of limitation associated with the practical constraints I've identified above. Usu-

---

[29]Edmund Husserl, *Philosophy of Arithmetic*, trans. with and introduction by Dallas Willard (Dordrecht: Kluwer, 2003), 202n.7. It does imply a particular conception of mathematical activity: Leibniz, for example, would have strongly disagreed.

ally, pragmatists and naturalists accomplish this by denying that numbers have any nature "in themselves" at all.

There *are* strong, good reasons for taking the attitude that our psychological limitations should play no role in our conception of number. After all, what numbers *do* is precisely to abstract from these psychological limitations! When we introduce new notation systems, we are building conceptual machinery precisely to *overcome* these limitations. And numbers, viewed as abstract, conceptual objects, are powerful precisely because (and even, we might argue, to the extent that) they permit us to leave these limitations behind. A sensitive proponent of this line of thought might even say: of course we don't ever leave these psychological limitations behind in an absolute sense, but the whole point of the development of an abstract conception of number is to overcome them as much as possible. The growing abstraction in our conception of number, and the concomitant abstraction in the mathematical domain at large, is the consequence of this ambition.

I'm unhappy with both sides in this philosophical standoff, but I'm not particularly interested in jumping in the ring and joining the fray. I think both sides are missing something so fundamental that, instead, I want to "brook" this debate by cultivating the notion of the parafinite. Although the concept of the parafinite can be made fairly precise, it takes a lot of work to do so, and in the present volume the concept will have to stay somewhat woolly. Rather than learning about it through a series of definitions which would require a lot of background in the foundations of mathematics, instead I'll try to help the reader acquire some sense of the parafinite much in the way we learn new words when we pick them up in conversation: by context, which is the way we all begin to learn (and, for the most part, how we continue to learn) language in the first place (a good part of what makes reading *Finnegans Wake* difficult is that we lack the *context* for doing so). In terms of the examples above, we're already in a position to appreciate what might be called the "practical parafinite." In the grocery store example, twenty is "just too large" to perform that many individual transactions. So we could say that in this practical context, twenty is a "parafinite" number.

This example illustrates another issue. One is not a parafinite number in this context: it's the norm. Is two already a parafinite number of grocery store transactions? We could debate about this, arguing pro and con (my experience seems to suggest the answer is 'yes'). And while in principle we could debate about the number twenty, it's pretty clear that twenty is already a parafinite number of transactions in the grocery store context. This exemplifies a salient feature of what I'm calling the parafinite: where it begins will often be vague. Here's another example (due, originally, to Yessenin-Volpin): consider the number of heartbeats in your childhood. If you're in your fifties, as I now am, you're no longer a child, but is there a last heartbeat in your childhood? It would seem pretty arbitrary, ridiculous even, to pick out a particular heartbeat and nominate it "the last heartbeat of my childhood." But there are clearly some events which are heartbeats in my childhood, and other events which are not. Then there are others in the middle for which it's not so clear. Some kind of a scale is being established here, but where one passes over from one part to another seems inherently vague.

A first sense of the parafinite, then, is that the parafinite is what outstrips a particular scale. Sometimes it will be clear what the first thing is that's beyond a scale limit, but sometimes it won't. What seems almost universally denied is that *there is any sense of scale intrinsic to the notion of number itself*. In this book, I won't argue against this directly by insisting that there are intrinsically large numbers, but I'll go ahead and admit my excitement for the idea and also my skepticism that the idea should simply be rejected out of hand.

For now, let's just register some implications of *denying* this strong version of the thesis. On this (standard) view, one, two and three are not in any intrinsic sense small numbers. There is not any sense in which $g_2(2,5)$, or even, say, $g_2(100,1000)$ is a large number. And despite the fact that we talk about large and small numbers all the time, on this view there are only "larger" and "smaller" numbers. Further, every number is both "larger" *and* "smaller." Three, for example, is larger than two and smaller than four.[30] This exemplifies the point Socrates makes in Plato's *Republic* when he calls such properties as longer and shorter (he's dealing with magnitudes rather than quantities) "summoners." In this passage, Socrates starts off with an example of comparative largeness and smallness in the domain of perception. Taking three fingers, call them small, medium and large, we perceive the medium finger to be short relative to the longer one and long relative to the shorter one. In a sense, it is both short and long! For Plato, this is a contradiction, and it can only be "resolved" by moving from the realm of perception to the realm of ideas, in which we separate out the ideas of shortness and longness from particular things, which stand in the relation of shortness to some things and longness to others. Socrates calls the middle finger a "summoner," since it summons us to the understanding of the short and long. When we look at a "summoner," like a finger, say, we see it as *both* short and long. In the summoner the two, the short and the long, are mixed together as one. In this sense, separating out the ideas of short and long from the perception of the summoner is a process of separating the "two" from the "one." Calculation and the understanding come to our aid in this process:

> Then it's likely that in such cases the soul, summoning calculation and understanding, first tries to determine whether each of the things announced to it is one or two.[31]

---

[30] Actually, as we'll see later, in a sense for the ancient Greeks, two is the smallest number.

[31] Plato, *Republic*, trans. G.M.A. Grube revised by C.D.C. Reeve (Indi-

Sight sees the middle finger as both short and long, but the understanding is required to separate the two.

Numbers and magnitudes are in some sense the most basic examples of summoners, for in this domain we are always dealing with things which possess contradictory properties like "large" and "small." For Plato, mathematics played a very important role in the "transition" from the domain of what we perceive to the domain of the "separated" ideas. But Plato's description of summoners also demonstrates that he already stands under the long shadow of the parafinite. For while numbers and magnitudes are in the domain of "mathematicals," largeness and smallness are not: they have already "transcended" mathematics, as they are located in the domain of "ideas." To be sure, this is rather a simplification of the role of mathematics in Plato's system, but the point is central. In fact, it leads to all sorts of wild speculations, both in Plato's own work, and among his followers, about "ideal numbers" that themselves transcend the domain of the sensible and so stand in direct relation to ideas like "the large" and "the small." The debates about these ideal numbers, throughout the philosophical tradition, have been among the thorniest and most interesting in the Platonic tradition of philosophy.[32] Here, I only wish to pose a question: since such ideal numbers stand in direct relation to ideas like "the large" and "the small," wouldn't they have to be intrinsically scaled? That is, wouldn't such ideal numbers have to come along with a scale, i.e. a fixed relation to the ideas of "small" and "large"? Plato never commits on this issue, but it seems like an obvious conclusion to draw from the idea of an ideal number. (Or maybe they come along with an idea of scale as such?!)

Plato's discussion of summoners is as important for the "prob-

---

anapolis: Hackett, 1992), 196 (524b).

[32]See the discussion in Jacob Klein, *Greek Mathematical Thought and the Origin of Algebra*, trans. Eva Brann (New York: Dover, 1992), 79ff. I return to Klein's work, but not specifically to this debate, in Chapter One.

lem of the parafinite" as the much more famous passage in the *Republic* about the ascent from the cave is for the Western philosophical conception of truth. It is a kind of foundational text for the "problem of the parafinite." In its own way, "summoning" is itself a more basic, even more fundamental, example of this later ascent toward the light of the Good: it was not for nothing that Plato had written over the doors of his academy:

LET NO ONE UNTUTORED IN GEOMETRY ENTER HERE.[33]

Above, we've seen what it would mean to say that a particular number is "unreasonably" large, hence parafinite, in a given practical context. Plato's ideal numbers might in some sense be "intrinsically" small or large, or at least stand in some particular relation to the ideas of the large and the small. But coming back to the numbers as we usually consider them, what could possibly serve as evidence that a particular number, or indeed that any number whatsoever, is parafinite independent of any practical use to which it's put? Much as I like the idea, I have to admit that such evidence isn't readily forthcoming. But in terms of this dilemma, for the first time, I can make the plan of this book clear. What I want to do is show historically what sort of problems arise in the *absence* of the idea of the parafinite, that is, the intrinsically large. In a few of these historical cases, thinkers even come close to considering something like this idea– in Galileo, for example, the

---

[33] Actually, the evidence for this ascription comes rather late: it can only be traced back as far as the fourth century CE and so the report may indeed be apocryphal. But in any case Plato abundantly emphasizes the role of geometry in the ideal curriculum for the philosopher as presented in the *Republic*. For a discussion of these issues see David Fowler, *The Mathematics of Plato's Academy: A New Reconstruction*, 2nd ed. (Oxford: Clarendon Press, 1999), 199-204, and also the review by I. Grattan-Guinness in *Mathematical Gazette* **84** 499, 165-6.

notion of a "quantity between the finite and the infinite." But generally speaking, the idea of the parafinite is avoided, and it is only in the context of really radical investigations of the foundations of mathematics that we encounter glimpses of this idea, "shadows" which the concept of the parafinite casts down.

It might seem extremely strange to the philosophically uninitiated to argue for an idea on the basis of this sort of indirect evidence: that we see the history of the concept of number brush up against the idea of largeness only time and again to drift back away from it, with the encounters becoming ever more aggravated. In fact, the general explanatory strategy at work here is one strongly promoted by the famous Enlightenment philosopher Kant. In the *Critique of Pure Reason*, Kant argues that if we can make a hypothesis which resolves hitherto unresolvable contradictions, then this should serve as strong evidence in favor of the hypothesis, *even if there is no possibility of verifying the hypothesis directly.*[34] This sort of hypothesis is referred to as "transcendental," since it transcends what we can verify directly. In the *Critique*, Kant shows how various philosophers have held contradictory positions, where the contradictions can be resolved by adopting his transcendental hypothesis. In this sense, Kant's method is also implicitly historical, since he identifies the contradictions to be resolved in terms of positions drawn from the history of philosophy.

In this book, I will not be looking at such contradictions, but I will be looking at the history of mathematics, science and philosophy in order to try to show up certain limitations in the traditional capacity to think about mathematical quantity. And I will be intimating– this will be the force of the cumulative historical presentation– that by adopting the hypothesis of the parafinite– of an intrinsic structure of largeness embodied in the counting numbers– these limitations can be overcome. We see a first in-

---

[34]Immanuel Kant, *Critique of Pure Reason*, trans. Werner Pluhar with an intro. by Patricia Kitcher (Indianapolis: Hackett, 1996), 21 (B xvi).

road in Husserl's early attitude toward number. It is not just that we, as human beings, have certain particular limitations: in our ability to perceive, in our ability to remember, in our ability to combine. It is, rather, that the very essence of numbers themselves is predicated, in their "operability," on such limitations. Perhaps we could imagine quantities of things independently of these limitations, but we can't imagine mathematical *operations* in the absence of these limitations, or such at any rate is my underlying claim. In the drama of the development of notions of quantity in the Western tradition, I plan to pay particular attention to the ways that these limitations inform our conceptions of numbers and the numerals which name them.

There is one obvious candidate for the role of the intrinsically large that I haven't yet acknowledged as such: the infinite. And, indeed, in Cantor's transfinite set theory we develop larger and larger "intrinsically large" numbers: namely, the infinite ones. To say that this has been an important development for modern mathematics is an understatement: mathematicians routinely acknowledge that in some sense modern mathematics would be "impossible" without Cantor's transfinite theory. The disadvantage, though, of thinking about these transfinite numbers as infinite is that it shields the finite domain from concerns about intrinsic scaling, and hence largeness. Very coarsely, the point can be put this way: if transfinite numbers are intrinsically large (because larger than every finite number), then there's no need for finite numbers to be intrinsically large. This deflects attention from issues of scaling in that domain that is traditionally taken to be finite. In fact, there are strong philosophical reasons to think that the distinction between the finite and the infinite is not so clear cut as it's often taken to be (I won't develop them in this book since they're rather involved, but I will give some references).[35] All of this points in the direction

---

[35]See, in particular, the discussion of Petr Vopěnka's Alternative Set Theory (AST) below.

of breaking down those "shields" against recognizing a domain of the parafinite, figuratively speaking, "between" the finite and the infinite.

There is another, weaker, version of the thesis that there is a sense of scale intrinsic to the notion of number itself, and this version of the thesis is something we will look at in more detail toward the end of the book. This is the idea that while there is not a fixed, particular sense of scale associated with the counting numbers, that there is a structure of scal*ing* associated with the natural numbers. This would be the structure that allows us to compare numbers and say that some are smaller (or larger) than others. Suitably formulated, this form of the thesis is so much weaker that it is in fact uncontroversial. But even in this weaker version, as we will see, there may be great value in *focusing* on this scaling structure, even if it is not particularly controversial once you see it. Putting things in the right light can often help change the way we ask new questions, and asking new questions can lead to major developments.

Although I take only a first, small step toward the parafinite in this book, the stakes are ultimately large. Talking recently to a friend, I mentioned the early twentieth century mathematician and philosopher Brouwer's attempt to make numbers intrinsically temporal. Brouwer's program was an attempt to face as frontally as possible a philosophical undercurrent of temporality that runs alongside the growth in the conception of number as abstract. Brouwer sought to wrest this temporal undercurrent from the shadows, bringing it integrally into the heart of the conception of number. For Brouwer, roughly, the number ten is the *process*, or *construction*, of counting from one to ten; even the number one is the process of counting "one." I told my friend that the project I am proposing is analogous to Brouwer's: Brouwer made numbers intrinsically temporal, and I want to make numbers intrinsically scaled. In this book, what I want to do is intimate an idea of scaled numbers by looking at the historical background of this concept. I hope that others will pick up on this idea of scaled numbers, perhaps even readers of this book. The mathematics of scale is still in its infancy, and a first step would be to think of numbers as always, in some sense, coming along with a scale (though this is a very crude first step, indeed). One on a scale of ten will be a *different number* than one on a scale of one hundred, or one on a scale of one million, or ... The idea is simple, but the consequences are dramatic– and we can always drop the associated scale when we don't need it. The reason for this reversal of orientation is that we are too tied in to our traditional conception of numbers as scale independent. Looking at the historical development of the conception of number we need to break away from, and seeing how it has caused problems associated with handling largeness, are very modest first steps in an enterprise that could ultimately be quite radical in its implications. If the rethinking is so fundamental, this may be a case where philosophy and mathematics need to work together. I hope so.

We begin, in Chapter One, with a consideration of quantity and magnitude, and the way these changed in the transition from ancient to modern mathematics. As we've already seen above, debates about the status of quantity and magnitude continued well into the nineteenth century context when Cantor established a kind of canonical divorce of quantity from any reliance on an appeal to magnitude. To understand the later debates about the infinitesimals that Cantor ultimately banished, and the appeal to the infinite that Cantor set up instead, we must start with Greek debates about quantity and magnitude. These lead naturally to what I will call a "fractional" approach among the ancients, which I will contrast with a "functional" approach that is quintessentially modern. These concepts will be our first building blocks for approaching the drama of the long shadow of the parafinite.

Navigation: A Reader's Guide

In one respect, this book resembles mathematics books more than it does books in the humanities (in all others it doesn't). This respect is that the chapters are not all at the same level of difficulty and that (therefore) multiple tracks through the book are possible. I've placed the Prologue before this Reader's Guide because I consider it a prerequisite for all of these tracks. Now that (I assume) you've read it, you may choose one of the following recommended tracks. But you may also use the suggested tracks as a guide for mixing and matching among them as you see fit. This guide permits the book to be used by readers with a variety of levels of mathematical and philosophical background: it is a multi-book rather than one chain-link narrative.

TRACK Z: Skip directly to Appendix E, "Four Principles of the Parafinite, with four quantum statistical analogues." Digest them and you're done: Zen Master!

TRACK A: Read each of the first chapters in each major division, omitting Section 1.4 from Chapter One. Also, omit the Appendices and the Envoi. This minimal track will still give you a strong picture of the overall design of the book. A Conclusion for Track A is in Section 5.4. Have a look at Appendix E.

TRACK B: Track A, adding back in Section 1.4 from Chapter One, plus some or all of the second chapters in each major division. Generally speaking, Chapter Two is easier than Chapter Four, which is easier than Chapter Six. So you might try Chapter Two and see how it feels: if it seems comfortable enough, give Chapter Four a try after reading Chapter Three, and if that feels comfortable try Chapter Six after Chapter Five. If any of these second chapters seem uncomfortable, you can safely skip over the later second chapters. Try Appendix A and Appendix B, but you may omit Appendix C or skim it lightly. Also, omit the Envoi or skim it lightly. A Conclusion for Track B is given in Section 6.5. Have a look at Appendix E.

TRACK C: Read the entire book (footnotes at your discretion!), including Appendices and the Envoi, which serves as a Conclusion for Track C. Bravo!

Scene One

From Ancient to Modern: Kleinian Themes, Netzian Narratives

Chapter 1

Divisions of Grandeur (Kleinian Themes)

> Encounter with the quantum has taught us, however, that we acquire our knowledge in bits; that the continuum is forever beyond our reach. Yet for daily work the concept of the continuum has been and will continue to be as indispensable for physics as it is for mathematics. In either field of endeavor, in any given enterprise, we can adopt the continuum and give up absolute logical rigor, or adopt rigor and give up the continuum, but we can't pursue both approaches at the same time in the same application.
>
> – John Archibald Wheeler,
> "Foreword" to Hermann Weyl,
> *The Continuum*, xii.

1.1 Introduction

In this chapter, I discuss two fundamental concepts, geared to two different routes, for thinking about the division of quantity: fractions and functions. Historically (and probably also conceptually), the consideration of aggregates, collections of units and so what we might call "multiplications" of quantity, has always been considered prior to divisions of quantity, and so the characterization of the small goes first by way of the large. In the context of various elaborations of the small, the concepts of fraction and function enter. On the face of it, it seems that a characterization in terms of fractions is conceptually simpler, and so more basic, than an approach by functions, and historically fractions appear long before the modern function concept. In this chapter, I will be focusing on two historical episodes and the transition between them to try to understand how fractions and functions enter in to the division of quantity: first, the classical Greek period, which culminates in the Euclidean theory of ratios, and the context of Leibniz's invention of the calculus, which is still predominantly fractional in orientation, but where the function concept serves as a kind of limit for certain features of his investigation. In the Greek period, I will be looking in particular at the brilliant and provocative account given by Jacob Klein in his book *Greek Mathematical Thought and the Origins of Algebra*.[1] This work is especially helpful as it establishes a framework for thinking about the development from ancient to modern conceptions. In the case of Leibniz, I will be relying primarily on the seminal paper by H. J. M. Bos, "Differentials, Higher-Order Differentials and the Derivative in the Leibnizian Calculus,"[2] appealing in particular to Sections 1 and 4, which treat respectively Leibniz's conception of mathematical quantity and his foundational studies of the Differential Calculus.

The chapter will conclude with an epilogue in which I suggest that, retrospectively, we may see that there is *something* in the classical Greek context that "goes proxy" for the function concept

---

[1] Jacob Klein, *Greek Mathematical Thought and the Origin of Algebra*, trans. Eva brann (Cambridge, MIT, 1968). Henceforth cited internally as GMT.

[2] H. J. M. Bos, "Differentials, Higher-Order Differentials and the Derivative in the Leibnizian Calculus," *Archive for History of Exact Science* **14**, 1-90; hereafter cited internally as DLC.

we only find developed later, in the modern period. This first chapter is focally about this classical Greek context, but I use later developments, and especially Leibnizian innovations, as points of comparison.

1.2 Kleinian Themes

To my mind, Jacob Klein's work remains the single most powerful attempt to provide a philosophical framework for understanding the historical development of Western mathematics, and in particular the historical evolution of the concept of quantity in the Western mathematical tradition. For our purposes here, it will be helpful to divide Klein's treatment of the ancient development schematically into three "subepisodes" which we may call the Pythagorean, Platonic, and Aristotelian episodes, respectively.

Pythagorean arithmetic works in terms of a "pebble presentation," which is not given explicitly by Klein but can be found presented in the work of Wilbur Knorr.[3] The Pythagoreans were able to depict the structure of quantity visually in terms of distributions of "pebbles," i.e. small point-like elements arranged in an array. The most basic distinction in the Pythagorean treatment of quantity is that between the even and the odd: this is the difference between a line-array that can be divided into two equal parts and one that cannot:

$$\bullet \mid \bullet \quad \text{versus} \quad \bullet \bullet \mid \bullet \quad \text{or} \quad \bullet \mid \bullet \bullet \, .$$

In two dimensional arrays, the most basic are given by the triangle, the square, and the oblong rectangle, which represent the

---

[3]W. R. Knorr, *The Evolution of the Euclidean Elements* (Dordrecht: Reidel, 1975), 142-61. I first learned about Knorr's treatment from a preprint version of Mitchell Miller's paper, "Figure, Ratio, Form: Plato's Five Mathematical Studies." My discussion closely follows Miller's synopsis of the relevant points from Knorr.

numbers, the odd numbers, and the even numbers respectively:

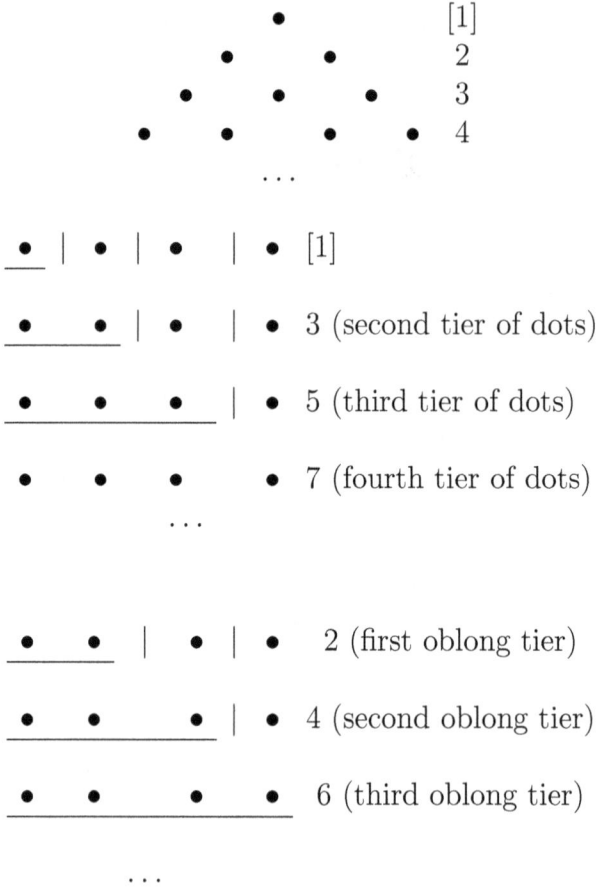

```
                    •           [1]
                •       •        2
            •       •       •    3
        •       •       •    •   4
                  . . .
```

```
 •  |  •  |  •   |  •   [1]

 •     •  |  •   |  •   3 (second tier of dots)

 •     •     •   |  •   5 (third tier of dots)

 •     •     •       •  7 (fourth tier of dots)
              . . .
```

```
 •     •  |  •  |  •     2 (first oblong tier)

 •     •     •  |  •     4 (second oblong tier)

 •     •     •     •     6 (third oblong tier)

              . . .
```

I have bracketed the quantity '1' referring to the first level of the first and second diagram, since unity had a special status among the Greeks as the "source" of quantity, and was not considered a quantity itself in the most straightforward sense. Indeed, the first two diagrams graphically depict the way in which unity "generates" quantity from the two figures respectively: first counting quantity and then odd quantities. That these two diagrams share this same generation from unity suggests one reason why the Pythagoreans

would take "the odd" as the consummate category of number rather
than "the even," which generates instead from "two-ity." According to Aristotle's report (*Metaphysics* M 6, 1080b20f., cited GMT,
67), the Pythagoreans identified the odd with "limit" and the even
with the "unlimited." Given the strong commitment to definiteness which is shared by all ancient mathematics, the association of
the odd with that which is limited indicates a definite priority of
the odd over the even. In the graphic depictions given above, we
may *see* this in terms of the way the associated pebble presentation
terminates at the top (either the apex of the triangle or the upper
left corner of the square) in unity, the generative source of number.
This is a "perfection" which the oblong representation of the even
does not share, and the oblong is also quite *apparently* defective in
the asymmetry between the sides of the oblong tiers. In fact, the
most important single diagram for the Pythagorean tradition is the
triangular diagram cut off after the fourth level (as above), which
organizes ten pebbles in terms of the first four "tier numbers." We
will return to this triangle below in the discussion of Plato as well.

The historical association of the Pythagorean tradition with
various forms of religious mysticism is undeniable, but Klein
stresses instead that the Pythagorean doctrine that "all is number" should be understood in terms of a sensory world-orientation.[4]
In our experience, we find the world to be structured in terms of
the divisions of bodies one from another, and in many cases this
structure is exemplified in terms of the basic phenomenon of aggregation. According to Klein, when the Pythagoreans said that "all
is number," what they were registering was a commitment to the
reality of this structure and its grounding in the basic phenomenon
of aggregation. This leads to a conception of number which be-

---

[4]A philosophically useful point of comparison is Oskar Becker's essay on
Pythagoreanism, "Die Aktualität des pythagoreischen Gedankens," in *Dasein
und Dawesen: Gesammelte philosophische Aufsätze* (Pfullingen: Neske, 1963),
127-56.

comes more philosophically explicit in Plato, but which we should already understand as associated with the Pythagorean tradition: for the Greeks, an *arithmos*, a number, is always a number *of something*, and, with an important potential exception, things *in this world*. That is to say, the Greeks do not speak, or think, of "the number four," say, in the way that we moderns (are taught to) do; they speak of "a four of apples" or "a four of stones," or, in the limiting case (as we will see) "a four of pure units." But even in this last case, the Greeks are not speaking of "the number four," since a four of pure units is only one instance of a four, and a four of apples or of stones would equally count as others. As Aristotle says, "It is quite rightly said that the *number* of sheep and dogs is the *same* if each is *equal* to the other, but the '*decad*' is *not the same* [in these cases], nor are the ten [sheep and dogs] the same [ten things], just as the equilateral and the scalene triangle are not the same triangles, although their *figure* is the same, since both are '*triangles*'" (*Physics* 224a2ff., cited GMT, 47).

Given the Greek commitment to the concreteness of numbers, it is relatively clear (though not uncomplicated) that the Greeks would tend to identify fractions with proportions: a fifth, for example, would be one among five, or two among ten. However, while we would typically *identify* the fractions 1/5 and 2/10, for the Greeks these would be two *different* fractions since the units entering in are different. In particular, this means that the concept of fraction is derivative from the concept of *arithmos*. This is not terribly difficult to appreciate from a modern perspective. What is considerably more difficult to appreciate from a modern perspective is that this means that a fraction *is not a number*. A fraction is not a number, an *arithmos*, because it is not a definite collection of things. A fraction, unlike an arithmos, is intrinsically relational. It is, in modern parlance, a ratio or proportion. This means that fractions do not have the sort of *ontological* status that *arithmoi*, "numbers" in the Greek sense, do.

8

Much of the development of the Western mathematical tradition has followed along the lines of a constant expansion of the number concept, beginning with Greek *arithmoi*, which correspond to but are conceptually distinct from our counting numbers (beginning with two, however, since an arithmos is an *aggregate*), and going on to assimilate zero, the negative numbers, fractions, irrational numbers and ultimately even so-called "imaginary" and "complex" numbers. Each of these stages has required a fundamental rethinking of the concept of number, and each has led in turn to major innovations and mathematical achievements. Along the way, however, the concept of number has become increasingly less concrete or, as we might say, less "intuitive." Nowadays, my six year old daughter reports from her school lessons that zero is a "symbol," that it is not a number in the way that other counting numbers are. That she discusses this explicitly, if somewhat incoherently, is an amazement to me. On the one hand, it strikes me as a positive pedagogical state of affairs that these issues are not swept aside or hidden; on the other, I find it indicative of the "jump starting" we must now do to bring our children "up to speed" with an increasingly sophisticated conception of numerical quantity: pretty soon six year olds are going to need to be learning about bits and bytes to get where they "need" to be by high school. The question "where will it all end?" expresses, at least to me, a mixture of exhilaration, dizziness, giddiness and ultimate discomfort, a sort of roller-coaster ride through the modern mathematical experience.

In part, we can untangle, or at least appreciate, some of this sophistication by returning to the ancient Greeks and witnessing the care with which they attended to these basic concepts, forging a conception of number that ultimately made possible the compendia of Euclid and Diophantus. And as Klein emphasizes, it was the return of the early modern mathematicians to just these classical texts which was largely responsible (by way of creative misreading, however!) for the development of the modern algebraic tradition

9

in Vieta, Stevin, Wallis, Descartes and others in the sixteenth and seventeenth centuries.

It is time to turn to the historical passage through the three ancient stages of numbers and fractions I mentioned above: Pythagorean, Platonic and Aristotelian. As we will see, it was in considerable part just the question of the status of fractions which drove this development. We have already seen the conception of fraction as ratio or proportion of units in the Pythagorean tradition. In Plato, this tradition is extended, but with the development first of a conception of aggregates of "pure" units, and then the conjecture of a rather esoteric doctrine of "eidetic" (or formal) numbers. Why did Plato find a need to push in the direction of form?

The answer to this question is, of course, complicated, with many philosophical dimensions which we will neglect here. But according to Klein, *the primary motivation for this development was a mathematical one.* If Klein is right, the consequences for Plato's thought are dramatic.

Arithmoi, as aggregates, are collections of things. But what are *collections*? If each thing has a unity which makes it one thing, what makes a collection of such things be one thing, i.e. a collection? It seems, in fact, that what makes it be a collection is that it is *not* one thing! For a collection to be an aggregate, it must be a multitude and *not* one thing. (It was, indeed, just for this same reason, that the Greeks could not consider one to be an *arithmos,* for it was *one*). That is, 'things' and 'multitudes' are mutually distinct categories. This may seem like "merely a matter of semantics," as we like to say today when we wish to dismiss something, but it is just such "semantic matters" that determine what will count as numbers and how we will think about them.

To put the point another way, and one that may be more comfortable in a modern mathematical context, what is it that mathematicians consider? Do they consider pigs and cows and stones and apples? No, they consider numbers. But what *are* numbers? There is a joke (though only some mathematicians seem to find it funny) that mathematicians are "formalists during the week and Platonists on the weekend." This means that when they are "at work," busy proving theorems and grading papers, they take mathematics to be an activity involving the formal manipulation of symbols which are "meaningless." As David Hilbert, who developed the modern doctrine of formalism, put it, whether you interpret mathematical statements about points and lines to be about beer steins and coffee mugs or any other collections of things is utterly irrelevant.[5]

But "on the weekends" mathematicians will (sometimes) admit that they believe in a "Platonic realm," that is, a sense in which mathematics expresses truths about a non-temporal reality that transcends and is more perfect than the world of becoming which we inhabit. If this weren't the case, if only formalism were true, then mathematicians would be nothing other than glorified, though admittedly very-high level, accountants, keeping the books. And surely not all mathematicians believe *that*. But this commitment to a transcendent Platonic reality, Plato's idea that thinking about number is a "learning matter," a *mathema*, capable of "hauling us toward being" (*Republic* 524E; 523A, cited GMT, 51) is, philosophically, an expensive one. It comes at the price of committing to a further reality beyond what is immediately, sensibly available. And just this commitment posed problems for the idea of fractions.

---

[5]In the *philosophical* sense of Hilbert's program, working mathematicians are not formalists during the work week or otherwise. Philosophically, Hilbert was committed to a grounding of the infinite in the finite that would essentially have reduced infinitary to finitary considerations. Even when mathematicians work in a "purely formal way," they are not (I claim) committing (either in principle or in practice) to such a reduction of the infinite to the finite. Further, it is generally thought that Gödel's incompleteness results "showed" that this program of philosophical reduction was untenable. But such matters are complicated.

Plato, not too surprisingly, will turn the tables and say that it is just this transcendental numerical reality which accounts for our ability to work as accountants, which is to say, to count and calculate "in practice." Behind this claim lies a whole host of tricky conceptual distinctions which Klein teases out– distinctions between arithmetic and calculation, between theory and practice, and between analysis and synthesis– but which I will largely leave aside here. Consequently, the account I give of Plato's contribution to this particular episode in the development of the concept of number must be taken as schematic in the extreme.

For our purposes, let's grant Plato his point about the need to appeal to pure units and, hence, pure *arithmoi* as aggregates of these pure units, so that we may ask what follows from this commitment. It's worth remarking, too, as Klein himself does (GMT, 51), that it is in terms of just such units that the classical Greek tradition (Euclid, Nicomachus, and others) defines *arithmos*.

First, on the basis of this conception of pure unit, what has come to be called an *iterative* conception of number can be developed, and it is explicitly promoted in such classical sources as Theon, Domninus and Nicomachus. As Domninus puts it, "the whole realm of number is a progress from the unit to the infinite by means of the excess of one unit" (cited, GMT, 52). Such a concept can proceed to infinity precisely because it is not dependent on the givenness of any other (worldly) things such as cows or stones, of which there is a limited quantity. Of the iterating of units there is no end.... But since we now are in possession of an unlimited progression of *arithmoi*, the problem of their organization recurs with increasing force. How are we to "organize" the various arithmoi of pure units? Here, Plato and many of those following him make recourse to the type of divisions proposed earlier by the Pythagoreans: as a "highest" division, into the even and the odd, and following on to other "lower specifications," beginning with properties like "even times even," "odd times even," and "odd times odd" (which as Klein notes, are not understood to be mutually exclusive: consider the number 36, which satisfies the first two properties).

This, however, does not yet resolve the problem which, according to Klein, most vexed Plato: if an aggregate is such precisely because it is *not* a unity, what reality can numbers have?[6] To answer this question, Plato developed his doctrine of eidetic (formal) numbers. Behind every seven– whether of stones, rabbits or even pure units– lay the eidetic number 7, a formal reality and as such a unity.[7] Following Pythagorean doctrine, Plato restricted his account of eidetic numbers to the first ten numbers, represented in the basic triangle of four levels I have depicted above. But *unlike* the Pythagoreans, Plato made the numbers *separable* from our worldly reality by grounding number in the appeal to pure units and the eidetic numbers which served as the formal guarantor for pure *arithmoi*, i.e. collections of pure units. This demand of separability arises jointly from the nature of pure units and the consideration of what it means to be a unit. As a unity, a unit is indivisible: a unit divided is no longer a unit. Here Plato's position follows the mathematicians, or so he at least claims in the *Republic* when he has Socrates remark that mathematicians would laugh at anyone who would attempt to divide the unit "and would not allow it, but whenever you were turning it into small change, they would multiply it, taking care lest the one should ever appear as not one, but

---

[6]For the analogous problem on the dianoetic level, see *Philebus* 130E-131E and GMT, 80: "how is it possible that one idea in its unity and wholeness is "distributed over" the many things which partake in it?" The eidetic numbers do not by themselves resolve this problem for *arithmoi*, but they do make it possible to pose it (and resolve it) in an analogous way.

[7]Potentially, this way of thinking about the structure of numbers suggests a very different way of "organizing" the numerical domain, not in terms of "properties" like odd and even but in terms of the individual eidetic numbers themselves. But Plato's doctrine is speculative and underdeveloped in the dialogues, and so extrapolation is risky.

as many parts" (*Republic* 525E, cited GMT, 39). Hence fractions would be understood by mathematicians as proportions of *arithmoi* and not directly as *arithmoi* themselves. But as Klein points out, the *pure* unit is the one which is indivisible "simply." As Theon puts it, this unit, "since it is noetic [i.e. pure], is indivisible while the one, inasmuch as it is sensible, can be cut infinitely" (cited, GMT, 41). The pure unit provides the limit of all partitioning, which will "stop at once" with this unit (Theon, cited GMT, 42). In short, there is no division of the unit.

Such is the "reality" of Platonic "arithmetic," but Plato also speaks of a *logistike*, or what we might call an art of calculation. Although, as Klein points out, Plato leaves underdetermined whether the "material" (*hyle*) of calculation is sensible or noetic, in practice the approach of the mathematicians was to view the underlying unit as sensible precisely because this was the only kind susceptible of partitioning (GMT, discussing Olympiodorus and the *Gorgias* scholiast, 60). There thus remained a contradiction between Platonic foundations and the actual practice of calculation.

This impasse was resolved by Aristotle, in our third subepisode, by arguing for the primacy but not the separability of the unit. In this way, the unit could be understood as a measure relative to which counting was performed. Thus what was chosen as unit could be picked in the most practical way for whatever mathematical problem lay at hand. Aristotle maintained the Platonic concern for unity by insisting that the unit was separable in reflection but not in reality, and thus he initiated a long tradition of theories of "abstraction" (*ex hyphaireseos*, cf. GMT, 133). Form, rather than being ontologically separable, is something which we may separately *consider*, and this holds in particular for number. Most importantly, Aristotle's "relativization" of the notion of unit which this abstraction theory made possible allowed the unit of measure to be changed *in the middle of a calculation*, so that a "floating" conception of *arithmos* was made possible that accommodated the

14

technical demands of fractional calculations. In the terms I have used above, Aristotle is developing a rudimentary mechanism for shift in scale.

We could add as a fourth episode what might be called "the Euclidean consolidation." Euclid was not himself so much an original mathematician as a compiler and canonizer of Greek mathematics. The historical impact of this canonization is undeniable, and we should not begrudge him the recognition such work deserves. But so far as our story is concerned, what is most important in Euclid's *Elements* is his presentation of the Eudoxian theory of proportions. This is material whose historical place has been considered and evaluated extensively, but it is not central for Klein's project, and so I will not dwell on it here. Klein does not dwell on it presumably because he takes as his focal concern the *conceptual* relation between unit and fraction in Greek mathematics rather than a development *in extenso* of the theory of proportion. However, this should not be taken to imply that the latter is in any sense irrelevant to the former. Rather than try to speculate on what Klein would have said about Eudoxus' theory of proportion had he considered it in more detail, I will restrict myself to supplying references to some historical analyses which I would recommend to the reader interested in considering these developments in the theory of proportion.[8]

---

[8]The most stimulating and provocative account of the Eudoxian theory of proportions I have encountered is that in Howard Stein's essay, "Eudoxus and Dedekind: On the Ancient Greek Theory of Ratios and its Relation to Modern Mathematics," *Synthese* **84** (1990) 163-211. Stein responds to the work of Ian Mueller, *Philosophy of Mathematics and Deductive Structure in Euclid's Elements* (Cambridge, MA), 1981, where one may find references to earlier literature.

1.3 From fractions to functions

So far I have spoken only of fractions and not functions, and it is a general commonplace that the function concept only comes into the Western mathematical tradition in the modern period. Indeed, on can argue, and along Kleinian lines, that the concept of function does not even emerge at the beginning of this modern development, but only in the seventeenth, or perhaps even eighteenth, century, depending on how you mark it. In the next chapter, we will see that Reviel Netz identifies relevant connecting links in the development of algebraic thinking which Klein neglects, and which have a bearing on the emergence of the function concept, at least indirectly. Nonetheless, the emergence of the function concept is deeply connected with just those transformations which first established the early modern mathematical attitude, and so I would like to sketch, again very briefly, this trajectory as Klein presents it. Klein insists that, on the one hand, the decisive modern advance in mathematics occurred only when early modern mathematicians turned to a systematic and careful reading of the ancients, but on the other hand, he also insists that the modern advance was a product of creatively misunderstanding this ancient work, a "strong misreading," if you will.

Following in the phenomenological tradition of his teachers Husserl and Heidegger, Klein effectively appeals to the distinction between a "natural" and a "phenomenological" attitude in his remark that "in its sum-total Greek science represents the whole complex of those *"natural" cognitions* which are implied in a prescientific activity moving within the realm of opinion and supported by a preconceptual understanding of the world" (GMT, 119).[9] Klein's

---

[9]Klein makes this statement in full recognition of the work of Otto Neugebauer's on antique algebra situating, e.g. the work of Diophantus in a broader cultural context, see GMT, 127 and notes 141 and 145. For an extensive consideration of the Klein's philosophical orientation, see Burt C. Hopkins, *The Orign of the Logic of Symbolic Mathematics: Edmund Husserl and Jacob Klein*

assertion, then, is the *strong* one that the Greeks stood in a *special* relation to the natural attitude *despite* the existence and availability of other mathematical traditions (and in particular those appealing to pure units). This is a strong historical claim, and one which parallels the insistence on the Pythagorean rootedness in a natural orientation we've already seen above. In stark contrast to this Greek orientation, Klein sees the modern orientation as determined by its critical opposition to the prior conception of science (*scientia*) held among the scholastic thinkers of the late medieval period. Consequently, when the early moderns received Greek mathematical thought they viewed it specifically as an *antidote* to the science of the schoolmen, and this led them to read it in ways which would not have been possible in a situation where the natural attitude supplied the framing context.

What this made possible, in particular, was a development of ancient mathematics which cut it loose from its overarching commitment to the *determinate* nature of quantity. As Klein insists, whenever the classical Diophantus uses a symbol to refer to an unknown quantity in a mathematical problem, this quantity is understood simply as unknown, hence undetermined, *to us* (*pros hemas*) until the solution has been produced, whereas the quantity itself is assumed to be determinate, indeed determined by the problem itself, *ab initio* (e.g., GMT, 132, 140).

In the early modern tradition, Klein focuses in particular on Vieta. Vieta's reception of Diophantus is dominated by his overarching preoccupation with extracting a *method* from Diophantus. Making use of the work of Pappus, Vieta extracts from Diophantus a thoroughgoing interconvertibility of equations and proportions, so that he is able to declare that "a proportion can be called the construction of an equation, and an equation the solution of a proportion" (cited, GMT, 160). In the most simple-minded terms, which do not do justice to the complexity of Vieta's program, we

_____

(Bloomington: Indiana, 2011).

may nonetheless begin to understand his claim according to the following simple-minded example. If I seek a proportion, say, "3 is to 4 as 6 is to what?," then I can express the proportion by:

$$3 : 4 = 6 : x,$$

and this is interconvertible with (i.e. equivalent to) the equation:

$$\frac{3}{4} = \frac{6}{x}.$$

When we "solve for x" in the latter equation, we *establish* the sort of proportion expressed above it. Vieta's emphasis is on recognizing this interconvertibility as the specific mark of a "method of finding," or what Klein calls the "finding of finding." A method in this sense does not just find things, it finds *ways* of finding things. Since the goal is to produce a general method for the solution of such problems, *the drive is in the direction of a formal calculus in which no question about the determinate status of the variable x need arise.* Indeed, the drive to generality ultimately recommends that we think of x as a variable, hence indeterminate quantity. Although there is only one possible solution in the case given above, what variable quantities in equations typically serve to do is to express general (hence in the Greek sense not fully determinate) relations among quantities. In this new way of looking at things, the relations among the quantities involved should only be constrained by the conditions of the original problem (GMT, 164).

There are fascinating possibilities which Vieta's orientation opens that can only be mentioned briefly here. In particular, Vieta's permission of indeterminate quantities opens the possibility of linking arithmetic problems of the sort investigated by Diophantus with the domain which Pappus identified as geometric, thus extending the "method" beyond anything either of them was able to accomplish in a Greek context. The interconvertibility of proportion and equation leads on to an interconvertibility of theorem

18

and problem, which deemphasizes the role of particular mathematical truths and places a premium on *"the finding of 'correct finding'"* (GMT, 166). These developments put fractions, as symbolic quantities, on a par with those counting quantities that would earlier have qualified exclusively as *arithmoi*, and these latter are themselves converted into the status of symbolic quantities on a par with all others (GMT, 176). Because of the assimilation of arithmetic and geometry to a general "algebra," these steps even open the door to the consideration of irrational quantities, hence to a veritable "burgeoning" of the conception of number. "As soon as "general number" is conceived and represented in the medium of species as an "object" in itself, that is, *symbolically*, the modern concept of "number" is born" (GMT, 175). Ancient mathematics is understood retroactively in these symbolic terms, and "this reinterpretation has to this day remained the foundation of our understanding of ancient "arithmetic" and "logistic"" (GMT, 176). Despite Klein's efforts to the contrary, I believe this remains largely the case, and at least in the English speaking world, Klein's work has only recently begun to receive the attention it deserves.

1.4   A Leibnizian Episode

I want to finish the body of this chapter by showing how the new mathematical attitude cultivated by Vieta and others led rather quickly to the function concept. This "case study" will necessarily be somewhat more technical than the previous portions of this chapter, but the example will also help to make the issues more concrete. If you wish, you may omit this section without loss of continuity and proceed straight to Section 1.5.

We see various aspects of Klein's discussion mirrored in Leibniz's conception of quantity, which embraces both arithmetic and geometric magnitudes. For geometric magnitudes, a full conception of quantity did not emerge until the nineteenth century, since in particular the mulplication of geometric magnitudes available in the seventeenth century was limited in nature. Such multiplication typically involved the problem of the multiplication of "dimensions," since geometric quantities were thought of as possessing various dimensions: 0 for a point, 1 for a line, etc. The product of two lines, it seems, should then be of dimension 2, but Descartes found a way to avoid this issue by defining geometric products by way of proportions. Given two line segments a and b Descartes chose an arbitrary unit segment 1 and then defined the product of segments a and b to be a line c standing in the proportion:

$$1 : a :: b : c. \quad \text{(DLC, 7)}$$

Note that this uses proportions to "define" a quantity and so resembles the interconvertibility of equations and proportions proposed by Vieta. As Bos points out, this strategy was useful in the theory of equations in one unknown, but dimensional homogeneity remained a requisite of the analytic[10] theory of curves into the eighteenth century (DLC, 7). Further, this dimensional homogeneity makes it possible to avoid the stipulation of an arbitrary unit length, and so we can work "intrinsically," without the quantitative analogue of a coordinate system. Ultimately the issue of infinitesimal quantities will force the recourse to such arbitrary unit measures back into play in the analytic theory of curves.

The *term* 'function' enters Leibniz's analysis in terms of the consideration of quantity: he uses it to denote a *variable geometric quantity*. But this use of the *term* 'function' is not yet that of what we now recognize as the modern concept of function, which requires the stipulation of a relation between an independent (input) variable (or variables) and a dependent (output, or functional) variable. As Bos points out, Leibniz later agreed to Bernoulli's use of the term 'function' to denote "the powers of a variable 'or any

---

[10]Here 'analytic' refers to the treatment of curves using the tools of calculus (as it was then understood).

function in general' of a variable," and the term thus lost its geometric connotations "and became a concept connected with formulas rather than with figures" (DLC, 9). The cultivation of the concept of an *independent* variable is a finer matter, however, and will require further discussion. It is here that the conceptual "battle" between fractions and functions reaches its apex.

Leibniz's approach to the analytic theory of curves was largely proportional or fractional in the sense that it considered general ratios of various geometrical quantities to each other without singling out any one as of specific importance. Yet there were foundational problems, in particular, that show the limitations of a proportional approach and drive in the direction of the modern concept of a functional derivative. Although Leibniz investigated these issues in detail, though not without residual error and perhaps confusion, he did so in unpublished manuscripts that had no impact on the historical development of mathematics. Indeed, as Bos notes, it is a peculiar fact about the history of mathematics that essentially none of the major investigations on the foundations of the calculus (Leibniz, Niewentijt, Paris Academy debates, Berkeley) had any significant impact on the forward progress of mathematics in the first half of the eighteenth century (DLC, 54). Although, in my opinion and to my knowledge, no fully satisfactory account has been given of this state of affairs, one (standard) way to understand the situation is in terms of the need for precise concepts of limit and derivative to further the foundations of the calculus in a way which would promote mathematical productivity. Also in my opinion, this explanation needs to be balanced against the ongoing productivity of "informal" infinitesimal and differential methods which continue to be used up to the present time, particularly in the "application" of the calculus to physical disciplines (see DLC, 11n.17).

The "fractional" approach to the calculus favored by Leibniz allowed for an unconstrained specification of the "progression of variables," which means (roughly) that the infinitesimal progression of the various variable quantities could be specified in any way desirable to the solution of a particular geometrical problem. In functional terms, this would amount to something like parametrizing all the variables in turn by an *extra* "time" variable t and then letting the variable quantities move along the course of the time variable at different temporal rates. The "fractional" approach avoids the complication associated with the introduction of an extra variable, but at the expense of its failure to supply any rigorous characterization of the various partial derivatives, or rates of relative change, of the various changing quantities. Here we see a latter day version of the earlier competetition between "arithmetic" (*arithmetike*) and calculation (*logistike*) among the Greeks, and as Klein points out, with the increasingly symbolic construal of quantity, the practical art of problem-solving came to assume the fore over foundational questions (GMT, 184)– this, too, should be taken into account in considering the failure of the foundational investigations of Leibniz, Niewentijt, Berkeley and others to influence the direction of mathematics. (Just as the business of America is business, the business of modern mathematics is solving problems– which includes proving theorems.)

When we begin working with infinitesimal quantities in the context of such "a *symbolic* discipline whose ontological presuppositions are left unclarified" (GMT, 184), we are dealing with a situation where a dual source of ontological indeterminacy (coming both from the symbolic conception of number and from infinitesimals having no finite, i.e. determinate, ratio to finite (symbolic) quantities) rapidly leads to potential mathematical vacuity. Perhaps it was ultimately just this source of underdetermination that motivated Niewentijt's rejection of higher order infinitesimals (i.e. infinitesimals infinitesimal relative to other infinitesimals, and so on). In any case, such higher order infinitesimals are so ontologically indeterminate in Leibniz's calculus that symmetry breakings in scale length and/or the specification of an independent variable are/is required to resolve them. Effectively, this is done through the arbitrary choice of an infinitesimal unit. In this way, we see a first instance of the way scale-specification and the function concept are deeply connected.

As Bos points out, the derivative occurs only in terms of the ratio of ordinate y to subtangent σ, yet geometrically there is no reason why we should not understand the derivative as the ratio x/λ, where λ is the segment extending vertically from ordinate value y to the intersection with the tangent line at that point where the curve meets the ordinate value 0.[11] The choice of the former over the latter amounts to the specification of x as the independent variable. Furthermore, if we understand the derivative geometrically as the correlation of a ratio of two geometric quantities (ordinate y and subtangent σ) with a variable (which, unhelpfully but tellingly we also designate y), then there is no natural way of extending this to the operation of taking a higher-order derivative. To do this, we would need to interpret y/σ as a line segment, produce a new curve and then take the derivative of this curve with respect to the variable y. But understood as a ratio, y/σ is a real number, and in order to interpret this as a line segment a unit of magnitude needs to be specified. Thus, the concept of higher-order derivative has no *intrinsic* geometric significance, or as Bos puts it, "in a purely geometric context, higher-order derivatives are not uniquely defined" (DLC, 8).

A similar, but more trenchant version of this problem can be seen in Leibniz's attempt to prove the rules of his calculus on the basis of what he calls "the law of continuity." This law states that "if any continuous transition is proposed terminating in a certain limit, then it is possible to form a general reasoning, which

---

[11]See the diagram given at DLC, 8. The triangle with sides y and σ and the triangle with sides x (elevated to the ordinate level y) and λ are similar, yielding the equality of ratios.

covers also the final limit" (*Cum prodisset*, cited DLC, 56). In particular, this will allow Leibniz to interpret expressions involving quantities dx and dy in the limiting case that these quantities are equal to zero. From the perspective of our current concerns, what is interesting to note is the way that this requires Leibniz to introduce an external unit (see DLC, 57). If we include differential elements dx and dy and introduce an arbitrary external (finite) unit $\underline{dx}$, then for fixed x and y we may define $\underline{dy}$ by the proportion

$$\underline{dy} : \underline{dx} = dy : dx \qquad (1.1)$$

In the limit, Leibniz identifies the proportion $\underline{dy} : \underline{dx}$ as the proportion y/σ, by a *geometric* limiting argument which does *not* make use of the law of continuity– he argues that the tangent is the limiting case of the secant line as the difference of values goes to zero.

If $dx \neq 0$, then the ratio $\underline{dy} : \underline{dx}$ can be substituted for the ratio $dy : dx$. Now by the law of continuity, the role of the ratio $\underline{dy} : \underline{dx}$ can replace the ratio $dx : dy$ in the limiting case, because it has been independently shown to make sense in the limit (where it equals y/σ), and so we can use the former ratio to interpret the latter when $dx = 0$, in which case we think of dx and dy as "fictions." Proving the correctness of the rules of the calculus then amounts to showing that when $dx = 0$, dx and dy may be replaced by the finite quantities $\underline{dx}$ and $\underline{dy}$.

Bos convincingly argues that this procedure for demonstrating correctness naturally leads to the functional concepts of differential quotient and derivative. If we allow ourselves retroactive recourse to functional notation and set $y = f(x)$, then (1.1) asserts for

$dx \neq 0$ that[12]

$$dy = \frac{dy}{dx}$$
$$dx = \frac{f(x + dx) - f(x)}{dx} \cdot dx,$$

and then the argument that the tangent corresponds to the secant in the limit amounts to taking $dx = 0$, so that

$$\underline{dy}\,|_{\underline{dx}=0} = f'(x) \cdot \underline{dx}. \quad \text{(DLC, 59)}$$

Bos also argues explicitly that Leibniz's strategy leads naturally to the introduction of the concept of function. In proving the rules of the calculus (such as the product rule) according to the method previously described, Leibniz's choice of a fixed unit $\underline{dx}$ is tantamount to a specification of x as independent variable, and the other variables are understood as functions of this independent variable. In the context of the treatment of infinitesimal differentials this is equivalent to taking dx to be constant, which is to say that we allow an arithmetic progression in the variable x (DLC, 61). Doing so constrains the specification of the progression of variables and so limits the practical power of the calculus. Although this is required for foundational precision, there is not (yet) any reason to do this in practice. This "tradeoff" reflects the general complementarity of rigor and application that John Archibald Wheeler speaks of in the epigraph I have chosen for this chapter.

1.5   Epilogue: Imaging

I would like to end this chapter on a final, speculative point. It seems that there is no function concept in ancient mathematics.

---

[12]Bos writes Δx and Δy instead of dx and dy here, but this seems unnecessary to me. We simply must interpret the formula with $dx \neq 0$.

Is this correct? While in a sense I believe this to be accurate, it is not because the structure which is reflected by the concept of function is lacking in Greek mathematics, but this structure does remain unformalized. A pebble "diagram" is a model or depiction, an image of the way in which aggregation presents itself to us in the world. In German, the word for depiction, *Abbildung*, is the same one which is used for a mathematical function. This is only a hint, of course, but a suggestive one. What it suggests is that the notion of imaging is at the root of the notion of function, where the "output" is the *image* of the input, and that in Greek mathematics this is still an integral part of the number concept, not yet separated off as a mathematical concept on its own. The growth of mathematics is a growth in the articulation of mathematical concepts. This growth, like growth in the biological domain, does not happen without attrition. As Klein has pointed out, the modern mathematical development is achieved only at the expense of an ontological opacity in the foundations of the modern number concept, one that lies at the root of what his "philosophical grandfather," Edmund Husserl, would call "the crisis of the European sciences." Indeed, this opacity will come back to haunt twentieth century philosophies of mathematics, including Husserl's, in a major way, and this will be a principal focus in our Third Scene. We should not see this opacity *exclusively* as something implied by a conceptual acceleration at the beginning of the modern age, an inability of modern mathematics to "keep pace" with technological and technical developments. It *is* this, in part, and it would be a mistake to ignore this element of the modern transition. But perhaps more fundamentally, it is a transition obeying some as-yet-insufficiently-articulated version of Wheeler's complementarity principle, in which a tradeoff has been made (this time) in favor of power over rigor. We do not usually think of the modern mathematical orientation as lacking in rigor. But, as G. H. Hardy has remarked, it was the lack of a powerful *technique* before Newton and Leibniz, not a lack of rigor, which impeded the progress of mathematics.[13] With the advent of powerful technique in the generation of Newton and Leibniz, rigor takes a back seat, though this should not at all be taken to imply that questions of rigor were abandoned or viewed as insignificant.

Yet what the mathematics of the seventeenth century lacks in rigor it makes up in ambition. As a philosopher interested in the history of mathematics, what interests me is to articulate the nature of this transition: a transition, in particular, which made the modern concept of function available for the first time.

---

[13]G. H. Hardy, *Divergent Series* (Oxford: Clarendon, 1949), 13.

Chapter 2

From Problems to Equations: Archimedes and Beyond (Netzian Narratives)

2.1   Introduction: Framing Questions

As part of a trilogy dedicated to ancient Greek mathematics, Reviel Netz has extended the investigations of Klein concerning the transformation of mathematics from the ancient to the post-ancient world. Rather than seeing a transformation as one from ancient Greek to modern mathematics, Netz insists on the roots of this transformation in the practice of late antique mathematics itself and sees a full-blown transformation of approach already in the medieval Arabic tradition.[1] Taking Archimedes and his reception as focus, Netz devotes the second volume of his trilogy to the problem

---

[1] Reviel Netz, *The Transformation of Mathematics in the Early Mediterranean World: From Problems to Equations* (Cambridge: Cambridge, 2004); hereafter cited internally as TM.

of the transformation of the problem-oriented mathematics of the Hellenistic world through the reception of Archimedes first in the later Greek tradition and then among the medieval Arabs. While Netz does not contest Klein's identification of a transformation of mathematics, he does identify a different root for this transformation, locating the most relevant pivots in issues of mathematical style of presentation. This is a marked shift away from Klein's emphasis on conceptual content and the transformation, as Klein identifies it, associated with the understanding of the fundamental objects of mathematics itself. Hence, rather than identifying a shift from Greek *arithmos* via debates concerning the status of fractions to a "fully abstract" conception of number in the early modern period, as Klein does, Netz sees a transformation associated most fundamentally with mathematical *practices*, ranging from the Greek emphasis on problems to the Medieval Arabic emphasis on equations. The reconstruction of these practices proceeds by taking into focal consideration the *style* of expression in the manuscript tradition that has come down to us.

Since these explanations are largely, if not entirely, orthogonal–Klein focusing on the *content* of the mathematics at issue, and Netz focusing instead on the *form of presentation* and the underlying *practices* such forms of presentation represent– it is not immediately obvious that the explanations proffered by Klein and Netz are in fact even incompatible. And indeed Netz stresses a fundamental continuity underlying the investigations of Klein with his own. But the continuity Netz identifies is largely one of addressing a problem in what he will refer to as "cognitive history": how are we to understand the transformation from the Hellenistic context, in which, as he puts it, the mathematics is "first-order" in nature, to the medieval Arabic (and by implication also later) context(s) in which the mathematics is "second-order." In order to understand what this means we will have to understand how Netz draws this distinction (which seems to be of a conceptual order),

30

why he thinks it is an appropriate way to couch the problem that Klein antecedently identified as requiring explanation, and how he sees Klein, first, and himself, second, as offering explanations for this transformation. Even after transacting this business it will not be entirely clear, I think, how conflicting the explanations of Klein and Netz are, but it will certainly become clear that Netz advocates a different *orientation* for offering solutions to these and other problems.

Here is the passage in which Netz parses Klein's approach:

> Klein's approach went deeper than the forms of mathematics. For Klein, it was not merely that the ancients used diagrams while the moderns use symbols. To him, the very objects of mathematics were different. The ancients referred to objects, directly, so that their arithmetic (the case study Klein took) was a study of such objexts as "2," "3," "4," etc. The moderns, however, refer to symbols that only then, indirectly, refer to objects. Thus modern arithmetic is not about "2," "3," and "4," but about "k," "n," and "p," with all that follows for the forms of mathematics. Ancient mathematics (and science in general) was, according to Klein, based on a first-order ontology; modern mathematics (and science in general) is based on a second-order ontology. (TM, 5)

Netz finds fault with Klein not for adopting this distinction, which Netz will himself accept, but rather faults Klein's approach for being *abstract*: "Klein's study was conceived in the terms of an abstract history of ideas that left little room for persuasive historical explanations" (TM, 5). Netz then goes on to stress Klein's emphasis on "Plato's statements about mathematics" rather than the mathematics texts themselves, and in general his focusing on " 'the concept of number' " (TM, 5). Netz, for his part, doubts that "generalizations can be offered at the level of 'the Greek concept of ...'" (TM, 5).

This hesitation points back in a curious way to Netz's own characterization of Klein's thesis. For Klein's thesis is indeed that we find a different concept of number in Greek mathematics than we do in modern mathematics, and not just that the "same" numbers are used directly in Greek mathematics and indirectly in later mathematics by way of their symbolization. So Netz's characterization of Klein's program itself relies on a rejection of the fundamental claim which Klein is making, at least in the way that Klein is making it. The "first-order/second-order" distinction is really a kind of "analogue" of Klein's thesis once the idea that the Greeks and later mathematicians possessed different concepts of number is rejected. And perhaps ironically, given Netz's historical ambitions, the explicit language of "first versus second order" is really only something that gets cemented in the mathematical logic of the twentieth century once a systematic distinction between mathematics and metamathematics is made. It is, of course, a conceptual distinction in its own right.

Without entirely foreclosing their compatibility, these and other remarks he goes on to make tend to imply that Netz does see his explanation-orientation as incompatible with Klein's. The more delicate question which emerges is: does Netz see an interplay between the transformation of practices and the transformation of concepts, or does he see the practices as what mathematics is "really about" and mathematical concepts as somehow superstructural? To be sure, Netz never identifies his position in such radical terms as the latter horn would imply, at least so far as I can see, and yet his emphasis on practices as underlying mathematical activity is quite strong.

Since the practices under consideration are historical rather than contemporary– and this is necessary given that Netz's goal is to provide *historical* explanations!– Netz's commitment to the primacy of practices amounts to a particular set of commitments in the *reading* of textual evidence. We cannot go out and look and see what Greek mathematicians were doing– or even early twentieth century mathematicians! One thing that this primacy of practices could imply is that we give a special privilege to *reports* of what earlier mathematicians were doing as opposed to the actual mathematical texts that come down to us. But although there is some evidence of this sort about what Greek mathematicians did and how they worked, for the most part Netz spends his time looking at the mathematical texts themselves with an eye to what they can tell us about Greek mathematical practice. His project is therefore largely reconstructive in a strong sense, and in particular a sense which is just as strong (though perhaps not as sweeping) as one which would attempt to reconstruct "how the Greeks thought." In the first book of his trilogy, *The Shaping of Deduction in Greek Mathematics*, for example, Netz pays a great deal of attention to the way that Greek mathematical diagrams are labelled, even counting the number of times a line is referred to as AB versus BA, to see what this can tell us about the way such diagrams were used.[2]

This general methodological orientation leads Netz to place a great deal of emphasis on presentation and what we may (and Netz does) refer to broadly as "style." The result is that the *syntax* (in the broad sense of structural organization) is prioritized over the *semantics* (content) of the mathematical texts under investigation. Netz's first volume is most radical in this regard, to the extent that Netz is able to declare that the reader is not required to know (or do) any mathematics to read the book. In the "Preface" he says that he will "study form rather than content, partly because I see the study of form as a way into understanding the content. But this content – those discoveries and proofs made by Greek

---

[2]Reviel Netz, *The Shaping of Deduction in Greek Mathematics: A Study in Cognitive History* (Cambridge: Cambridge, 1999); hereafter cited internally as SD.

mathematicians – are [sic] both beautiful and seminal. If I say less about these achievements, it is because I have looked elsewhere, not because my appreciation of them is not as keen as it should be" (SD, xi). To be sure, in the second volume of his trilogy we certainly see that Netz possesses a keen appreciation of the content of particular mathematical problems in the Greek corpus, and in fact he chooses to focus on the reception-history of one of the most ambitious problems tackled by Archimedes. Since Netz licenses us to focus on practices, we might at least question the extent to which his own practices have to do with the question, which he himself raises on the same page of the "Preface" cited above, "Is there anything left to say?" (SD, xi). Netz's approach is certainly novel (as applied to ancient mathematics– it is actually an instance of the dominant approach in the contempoary "Sociology of Scientific Knowledge" tradition). If we are to take the priority of practices as fundamental in an eliminative sense, we need ask no further questions about the viability of Netz's approach: it has since been referred to as a "masterpiece" (David Sedley) and a canonical work of Science Studies (Bruno Latour). But if we take Netz at his word, the question of the extent to which the form of Greek mathematics illuminates its content remains, and it is on this question that I will focus in what follows. Although this is not Netz's focus per se, I hope it will constitute a compatible extension of what he has done. In particular, this will make it possible to address the question: what conclusions about the historical transformation of concepts of quantity can we extrapolate from Netz's narratives?

2.2 Archimedes' Problem and its Reception

2.2.1 Archimedes' Innovation

To address the problem of the relation of form and content in Netz's reading of Greek mathematics and its subsequent transformation, I will focus on the second volume of his trilogy, where these issues are most explicit. In this volume, Netz looks at the reception history of a geometrical problem, first found in Archimedes, as a locus for considering the transformation from a problem-oriented to an equation-oriented presentation. In Archimedes, the presentation is manifestly of a geometrical problem, though Netz will find places in the demonstration where Archimedes accedes to a more equation-oriented language or style. The next major steps in Netz's narrative involve solutions of the same problem given by Dionysodorus and then Diocles. All these proposed solutions are grouped by Netz under the heading of "The World of Geometrical Problems." Following this we turn to a presentation of a commentary on Archimedes by Eutocius, and here we have formally entered a new world: not that of addressing geometrical problems "directly," but rather of commenting on previously given solutions to geometrical problems. As such, Eutocius already inhabits the world of "second-order" mathematical textualism, and Netz sees the introduction of algebraic expressions, though still anomalous, as a consequence of this second-order manner of presentation. For Netz, the mathematical practice of "commentary" thus leads to a mathematical innovation, the equational presentation. And yet, as Netz notes, the innovation is rather paradoxical in that it is precisely a lack of ambition on Eutocius' part which leads to the most mathematically innovative aspect of his commentary. This lack of ambition is registered in the fact that the algebraic-equational aspect of Eutocius' presentation remains in a fundamental sense anomalous, in the sense that the equational aspect of Eutocius' presentation is not integrated into a larger algebraic practice. It is only in the work of a later mathematician of genius, the Arabic Omar Khayyam, that the algebraic-equational manner of presentation leads to a fundamentally algebraic *solution* to Archimedes' original problem. At this point we see, then, a full transformation from the Archimedean practice of geometrical problems to an algebraically orientated solution of such problems in terms of equations. This solution then takes its place inside a larger practice which we may (and Netz does) refer to as "Khayyam's algebra." Such is, at least as I reconstruct it, the broad trajectory of this Netzian narrative. Let's now look at the pieces of this story in somewhat more detail.

Netz focuses on the fourth problem from the *Second Book on the Sphere and the Cylinder*. This problem involves cutting a sphere by a plane in such a way that the volume of the sphere lying on one side of the intersecting plane stands in a given ratio to the volume of the sphere lying on the other side. The idea is to solve this problem not just for one particular ratio, but for any ratio that would be supplied in advance. The simplest case is for the ratio of 1 to 1, in which case the two parts into which we divide the sphere should be equal. In this case we can do this directly simply by cutting the sphere by any plane passing through its center. Notice that any plane passing through the center of the sphere will cut the sphere in a "great circle," i.e. a circle whose radius is the same as the radius of the sphere itself. In the case of a division of the sphere into two equal parts, the problem can also be reduced to the problem of dividing a great circle (but not the one specified above) into two equal parts. Take a great circle and a line passing through the center of the circle: this line cuts the great circle into two equal parts. Then, if we extend the line to a plane passing through the great circle through the line at right angles to the circle, this plane will cut the sphere into two equal parts.[3]

Already in the case of a ratio of 1 to 2 we cannot handle the problem so simply. This ratio requires us to slice the sphere by a plane into one part twice the size of the other, i.e. we must effect a division into a third of the sphere and two thirds of the sphere. But if we take a "great circle" on the sphere, i.e. any circle obtained by cutting the sphere through its center by a plane, we cannot simply divide this circle in the ratio of 1 to 2 by an intersecting line and get an equivalent ratio for the sphere. To solve the general problem, Archimedes sets it in terms of the ratios of associated cones, since cones of equal base stand in the ratio of their heights. These heights are lines, and so we have reduced the problem of the ratios of the sphere to one of the ratios of heights.[4]

Archimedes announces two solutions to the problem, an "analytic" one and a "synthetic" one. We will return to these terms later. In the manuscript, neither of these solutions is given, and the modern commentator Dijksterhuis conjectures that the solution must already have been lost by the time of Diocles, who gives a solution instead. Eutocius, however, thinks he has found the solution again, "in an apparently mutilated form,"[5] and provides a reconstruction. As we will see, then, there are questions about

---

[3]Actually, the condition that the plane be at right angles to the great circle is not really necessary, but it makes the visualization of the entire picture "cleaner." In this case the original great circle will be at right angles to the great circle made by the intersection of the plane through the sphere.

[4]The diagram as given in Netz is misleading, since it represents the cone $A P \Gamma$, obtained by rotating about the line $A\Gamma$, as not intersecting the section of the sphere bounded by $A B \Gamma$. But this would mean that the cone has greater volume than the section of the sphere, when in fact the cone is to be constructed to have equal volume. The diagram is correctly given in Dijksterhuis, 193, but labelled in mirror inversion to Netz's diagram.

[5]E. J. Dijksterhuis, *Archimedes*, trans. C. Dikshoorn, with a new bibliographic essay by Wilbur R. Knorr (Princeton: Princeton, 1987), 195. Dijksterhuis' work first appeared in monograph and periodical form in Dutch between 1938 and 1944.

the provenance of the proposed "Archimedean solution," and in particular about where Eutocius' reconstruction ends and his commentary begins. In terms of the linear flow of Netz's own narrative, he defers questions about the provenance of the solution until later and simply presents "Archimedes' solution" first.

Before he begins dealing with Eutocius' reconstruction of Archimedes' solution, Netz draws some morals about the way Archimedes' frames the solution to the "general problem" in the text (which involves the formulation of the problem in terms of cones and more to boot, but I omit the details here). Archimedes reaches this problem, Netz says, not "like a schoolmaster, who tells us explicitly that a certain particular problem might also be conceived as a more general one. He reaches it, rather, like a conjurer. Having got us used all the while to thinking about a particular problem, suddenly he begins to talk about it as if it were *already* general ... We are not led into the general problem; we are surprised into it. This, I would say, is typical of Archimedes' style, trying to elicit from the audience the effect of awed surprise" (TM, 15). In fact, Netz will even go on later to speak of the creation of an "aura" in Archimedes' statements of and solutions to problems (TM, 114, 119, 125). Generality per se is not "an overriding goal," but rather finding "an original and elegant way of stating the problem (and then, of course, of solving it)" (TM, 57). As originality overrides generality, we will largely find generality in the service of originality. For this reason, when we find anticipations of equational presentation in Archimedes, it occurs as "spice" designed precisely to arouse a feeling of originality in the reader, rather than functioning as an end in itself.

In fact, Netz's story is more complicated than this, for he actually sees generality entering into the Archimedean presentation for two distinct reasons, only one of which we have described above. The second reason generality enters in has to do with the *complexity* of the result Archimedes presents, which becomes increasingly unwieldy in the absence of certain forms of generalization. (How we are to measure this counterfactual conditional 'in the absence of' remains only informally indicated, but the suggestion seems plausible enough.) The obvious question is: is this latter source of generalization a matter of style, or content, or both? And the equally obvious answer is that it seems to be inextricably a function of both. So here, Netz's story is not one *just* about form, but about content as well.[6] But by attending to Netz's *own* style, we will see that we can distinguish between these two strands in terms of their relative priority for Netz's narrative as well.

We can compare the two sources of algebraism that enter into Archimedes' solution by the way Netz characterizes them. First, neither one is motivated by a desire on Archimedes' part to seek generalization for its own sake (TM, 16). Second, the fundamental nature of Archimedes' problem is geometrical, framed in terms of geometric magnitudes (lines, planes, surfaces, and volumes), and any incursion of "algebra" registers a departure from this fundamental context. In this geometric context, the standard method of handling geometric magnitudes is to set them in proportion to each other, and any such proportion must respect the "type" of geometric magnitude involved. Thus a line may stand in a ratio to another line, a part of a plane to another plane magnitude, etc. Beyond this, a line may stand in a ratio to a plane magnitude *as* another line stands in ratio to another plane magnitude: in this

---

[6]In his book, *Synthetic Philosophy of Contemporary Mathematics* (trans. Zachary Luke Fraser, New York: Sequence Press, original publication in Spanish 2009), Fernando Zalamea has argued that twentieth-century philosophies of mathematics have often been vitiated by their insistent focus on "foundations" and elementary mathematics, which systematically skews their focus away from the problems that characterize mathematics "in the wild." Although Netz's second volume (historical rather than philosophical in nature) certainly is not guilty of this fault, nonetheless his emphasis on practice does tend to skew his focus in the way Zalamea indicates, and we will see that the strand of concerns associated with the complexity of Archimedes' proof takes a back seat to his concern with questions of Archimedes' "style."

case, too, the nature of the geometrical magnitude is respected. Furthermore, it is permissible to treat the product of a line by another line (*our* way of putting it) *if* we think of this in terms of the rectangle which two lines form when sharing a common terminus where they meet at a right angle: the magnitude may thus be understood geometrically as the area of the corresponding rectangle (the typical *Greek* way of putting it). Here, too, the "nature" of the various geometric magnitudes is respected. What registers a departure from such a geometric context into a more algebraic one is when we find manipulations which tend to depart, either by virtue of the locution involved or by implication, from the respect for the differences in the geometric magnitudes involved. It will help if we give an example where Netz finds such a departure to occur.

Just as it is permissible to handle the products of lines in terms of areas, so it is permissible to use rectangular solids to handle the product of areas and lines. But Netz finds a locution employed by Archimedes, "the {area}, on the {line}," which he will call 'the *epi*-locution', "meaning something like a multiplication of an area by a line" (TM, 19), in which this multiplication is not explicitly translated into the manipulation of solid volumes. Here Netz identifies a stylistic departure from a purely geometrical manner in Archimedes' use of such a locution. Moreover, according to Netz this is a departure which is not "forced" by the problem itself (this is our second source of generalization, and we will return to it shortly), but is rather a "free choice" on the part of Archimedes himself. As such, it pertains to the form of presentation in the most deliberate sense: its *style*. Netz's language is important here: "there is also a major way in which Archimedes' text, very surprisingly, makes a deliberate *choice* to deal with objects as if they were quantitative in nature. This choice, more than any other feature of Archimedes' text, points forwards towards a more algebraic understanding of the problem" (TM, 98, italics in original).

Netz's careful analysis of the expression yields the conclusion that the Greek locution he translates as "figure <multiplied>by line" is anomalous, since Greek mathematical language is (typically) geometric in nature and not "quantitative" in the way this translation of the expression implies. How do we account for such an anomalous usage? Here is Netz's explanation:

> In the Archimedean passages where our *epi* is used, Archimedes needs to discuss, effectively, prisms, and he needs to discuss them in the context of proportion manipulations and proportion manipulations only. The spatial presence is irrelevant and potentially disruptive, since the diagrams are already extremely complicated as they stand. Hence some terminological anomaly is *required*. (TM, 110, my emphasis)

Here a terminological anomaly is *required*, whereas previously Netz referred to Archimedes making surprising, deliberate *choices*. Are these two cases related, and if so, how?

Before attempting to address this question directly, let us return to the other sources which Netz identifies for nascently algebraic aspects of Archimedes' solution to his geometric problem. These are the sources, specifically, which Netz does *not* associate with surprising stylistic choices on Archimedes' part. Why not? Precisely because they are *compelled* in some sense by the solution at hand, and so in some sense are *forced* upon Archimedes. Since our current goal is to get some sense of what Netz means by a "free, deliberate choice" in the passage cited above, the specific nature of these junctures is less important than the way Netz characterizes them.

The first juncture is in fact to be found in Archimedes' framing of the problem itself. As Netz puts it, an "algebraic reading is especially tempting in this case, because the geometrical configuration comes with ready-made orthogonal coordinates . . . This is part

of a general orthogonal grid, within which the proof is conducted
...More importantly, the orthogonal grid is *necessary* to make the
hyperbola (which conserves rectangles, rather than general parallel-
ograms) conserve the equality between the areas $\Theta\Lambda = \Theta H$" (TM,
27, emphasis mine). In terms of this orthogonal grid, which reminds us of the (x,y) Cartesian coordinates of modern algebraized
geometry, it is easy (for us, that is) to define the parabola and hyperbola at issue, and even the cubic equation associated with the
problem, "in terms of multiplications and subtractions" (TM, 27).
But Netz stresses that "this is an exceptional situation. Generally,
Greek conic sections (and similar lines) are not embedded within a
system of lines comparable to our modern coordinates" (TM, 27).
Although Netz does not say so explicitly, the orthogonal "coordinates," if such they are, are completely internal to the problem at
hand and are certainly not conceived to have any status antecedent
to the problem in which they appear, as they would in the modern
case of a Cartesian coordinate *system.*

Netz identifies what makes this exceptional condition emerge
in terms of a specific complexity of the problem at hand: a hyperbola is constructed to equate certain rectangles (the details need
not concern us here) and a parabola is also introduced which "in-
tertwines" with the construction of the hyperbola in terms of the
same rectangles. These rectangles therefore comprise a "natural"
nascent system of "orthogonal coordinates." Netz summarizes the
situation:

> The parabola and the hyperbola each arise out of a spe-
> cific configurational need. In the case of the parabola,
> this is to simplify areas into lines. In the case of the hy-
> perbola, this is to align lines together along a diagonal.
> But because the constructions of the two conic sections
> are intertwined, they also become, incidentally, interre-
> lated. Besides each serving its own specific geometrical
> function, they also happen to be defined relative to the

> same lines so that one can – if one wishes – describe them as functionally interrelated. Heath did and, while Archimedes clearly did not intend any such functional understanding of the conic sections, it is interesting to see, once again, how the trace of a possible equation appears within Archimedes' geometrical problem. (TM, 28)

I would like to emphasize several things about this passage. First, so that there is no misunderstanding, note that the first appearance of the term 'function' in this passage means something simply like 'role', while the latter instance is rife with "quasi-algebraic" meaning. Heath, the early-twentieth century British historian of mathematics, is cast in the narrative role of mining these "traces of possible equations" for all they're worth.[7] Next, we should note that the appearance of 'incidentally' in the passage is ambiguous. Does it refer to the appearance of the hyperbola and the parabola in the solution to the problem, or to the fact that the *way* they are related in this solution makes the solution *particular*, so that the appearance of something like orthogonal coordinates here is *specific* to this particular problem? Netz has previously made the latter claim, but what he goes on to say seems to me to favor the first interpretation as the specific point he is making in this context. The reason this seems important to me is that adopting the first interpretation reinforces the idea that this context shows how there are, "in the abstract," opportunities for "algebraization" that Archimedes does *not* take up. This reinforces the idea that Archimedes' underlying *practice* is geometric in nature. This is borne out when Netz says later that "Greeks were perfectly capable of a quasi-algebraic treatment – but, in practice, they happened

---

[7]The mid-century work of Dijksterhuis is remarkably absent from Netz's volume: eleven page references to Heath versus none to Dijksterhius in the Index. In Netz's third volume, the tables are turned, but at an overall lower total: Dijksterhuis rates four page references, Heath one.

to minimize it. To account for that, then, we should understand the nature of their mathematical practice" (TM, 54-55). This, indeed, could be taken as the over-arching methodological thesis of Netz's entire trilogy, and illustrative of his ambivalent response to the work of Klein.

On this basis, there seem to be two broad cases in which Greek mathematicians would depart from a geometric treatment. One would be when a geometric treatment became sufficiently unwieldy that an alternative was effectively "required." The other would be when they had a very definite, particular reason for doing so. There is a kind of intersection of the two broad cases: an alternative is required, but a *specific* alternative is not forced. That is to say, in this "intersecting" case, *some* departure from a geometric treatment is forced, but not any *particular* one. Netz identifies just this as the case in Archimedes' use of the *epi* locution. For, on the one hand, Netz describes it as a surprising, deliberate and free choice on Archimedes' part. But, on the other hand, Netz says that it occurs in a context where "some terminological anomaly is required" (TM, 110). What remains to be understood, then, is why *this particular* terminological anomaly is so suprising *as opposed to other candidates*. Netz explains Archimedes' originality in these terms:

> The remarkable thing is that he chose to import an expression from the domain of arithmetical calculations; we would have expected him to import from a nearer domain.... The innovation consists in a new configuration of registers, a new intersection of contexts. (TM, 110-11)

More specifically (and more speculatively), Netz conjectures that Archimedes' resort to the distant land of arithmetic may have been occasioned by a desire to avoid the formulation in terms of geometrical prisms precisely because it would have stood in too close a linguistic "presence" to the related statement about cones (TM, 111).[8] This conjecture makes Archimedes' recourse a function of the linguistic-terminological (as opposed to mathematical) organization of the text and brings Netz's conjecture back into conjunction with the extended analysis of Greek geometric vocabulary he gave in the first volume of his trilogy, *The Shaping of Deduction in Greek Mathematics*. More broadly, it could be seen as an instance of Archimedes' desire to surprise by drawing material from distant mathematical registers; as such, Archimedes' power to surprise is a function of his heightened capacity for drawing analogically from disparate mathematical domains. Archimedean surprise is a function of the power of "disparity action."[9]

2.3 From Dionysodorus and Diocles to Eutocius and Khayyam

Since my concern will ultimately be with the interplay between the geometric and arithmetical in Archimedes, which returns in a major way at the beginning of the third volume in Netz's trilogy, I will treat the remaining trajectory of Archimedes' problem in its reception from Hellenistic through Arabic mathematics more cursorily. For readers eager to reach the punch-line, this section may be skipped.

---

[8]What I fail to understand about this suggestion is why the formulation in terms of prisms would be considered *anomalous* by Netz. Perhaps it would be anomalous precisely in the sense that it stands in too close a "competition" to the role played by cones in the proof. But since my aim is to provide a reconstruction of Netz's argument with a view to the issue at hand, I leave this aside. The references given in the text at the top of page 111 are manifestly incorrect: the sense requires that '*(28)' should refer back to what has previously been labelled *30, and '(27)' should refer back to (29).

[9]Here I draw on I. A. Richards' theory of metaphor; see his *The Philosophy of Rhetoric* (Oxford: Oxford, 1936).

In Dionysodorus' solution to Archimedes' problem, Netz discerns a "more quantitative understanding of its objects" (TM, 37), but this is not yet tantamount to an algebraic treatment of the problem, because "he simply nowhere presents an equation to be solved. This shows (again) that there is "a dialectical relation between the "geometrical" and the "abstract" (which we may even refer to as the "algebraic"). The two do not rule out each other: they coexist in complicated ways" (ibid.). Netz goes on to speculate about why Dionysodoros did not hit on Archimedes' solution (which was not available to him). The simple answer, of course, is that there are many ways to skin a cat, but Netz offers a suggestion about what may have motivated Dionysodorus' approach.[10] Since, Netz says, "in mathematical terms, it made little sense" for Dionysodorus in the *particular terms* of cutting the sphere (as he does), he suggests we look for an extramathematical reason. The reason Netz proposes is that if Dionysodorus solved the problem in the *general terms* framed by Archimedes (which *were* available to Dionysodorus), he could only be viewed as a "footnote to Archimedes," whereas if he solved the problem directly (and rigorously), the solution would stand out in a very strong sense as his own. Ironically, "in order to single himself out from Archimedes, he was forced to approach the problem not in any terms, but in the precise terms from which it arises in Archimedes' original problem" (TM, 38). In this way Dionysodorus goes Archimedes one better and gets the best of the bargain. This suggestion seems less than entirely persuasive to me, but after all, Netz only makes it as a sug-

---

[10]Netz's phrasing is logically overextended, and we must understand it rhetorically. He suggests we try to see why Dionysodorus *did not* solve the problem in the Archimedean manner, but he goes on to identify his suggestion as speculative ("We should *perhaps* look for extra-mathematical reasons: and an obvious one *may* be offered along the following lines" (TM, 37, my emphases). So we should take Netz as offering suggestions why Dionysodorus *might not* have produced Archimedes' solution. Insisting on this is important because the registers of *compulsion* and *free choice* are at issue.

gestion. What it does elicit is Netz's attentiveness to the generally *agonistic* context of Greek mathematical practice.

In the case of Diocles' solution, we find Netz employing once again the semantic register of being 'forced', but now in a different sense of the term's semantic range. Here what Netz underlines is the way in which Diocles' solution is *contrived*. Diocles understands the original problem "not in terms of *a single proportion of lines and areas* [as Archimedes and Dionysodurs had done], but in terms of *three proportions of lines*" (TM, 45). As Netz sees it, this sets the problem in "general terms," but ironically terms which are so contrived that one would only appeal to them should one have a *particular* problem in mind (like Archimedes' problem!) which satisfies them.

> While stated [by Diocles] in its general form, the problem is such that it has no meaning on its own. One is just never interested in getting three points fulfilling the three proportions stated by Diocles, unless one had a very special reason to do so. Archimedes' statement of the problem is very different in this respect: it is couched in such general and simple terms, that one can think of it as standing, so to speak, on its own feet. But Diocles' terms are so contrived, that they can be understood only as a thin disguise for a special configuration. (TM, 53–54)

Diocles' solution is unsatisfactory in the very terms favored by Greek mathematicians, for it lacks the naturality and freedom of expression favored by the Greek emphasis on originality.[11]

This completes Netz's survey of the Hellenistic tradition, and Netz next turns to the commentatorial tradition as represented

---

[11]Naturality and freedom of expression stand in dialectical tension; in general, Netz emphasizes freedom of expression, but here criticizes Diocles' generalization in terms of its naturality.

by Eutocius. In fact, the three proofs ascribed, respectively to Archimedes, Dionysodorus and Diocles come down to us in the commentary on Archimedes by Eutocius, working in the sixth century CE. With Eutocius, we enter what Netz calls a *deuteronomic* culture (TM, 121), one in which the practice involves not the first-order solution of geometric (or other) problems, but rather the second-order commentary on first-order solutions previously proposed. As mentioned previously, the irony is that Eutocius' originality is to be found precisely in his lack of ambition in the traditional Greek, mathematical sense. His innovations, and in particular his promotion of a functional and equational approach is an unintended by-product of his very different mathematical practice: "He stumbles across functions and equations, *without ever thinking about it*" (TM, 64). This happens in terms of the way Eutocius chooses to use numerical symbols to label diagrams.

Greek mathematics labelled numerals by letters of the alphabet, but the number of letters in the alphabet was insufficient for mathematical purposes, and so the Greeks introduced additional symbols to complete this labelling. What happens in Eutocius' presentation is that these additional numerical symbols happen to be used to label objects in the second half of his presentation of Archimedes' proof. In this way, "the second part of the proof declares its foreignness by its use of foreign symbolism. Once again, we see how deviations from an established practice are meaningful in themselves" (TM, 84). Since this was the part of the proof *not* supplied by Archimedes, and involved explicit generalization from Archimedes' proof, the generalization is strongly marked off by the novelty in symbolism.

Secondly, like Archimedes, Eutocius uses the "*epi*-locution," i.e. "object <multiplied>*by* (*epi*) object." But in Eutocius this appears, not, as in Archimedes, at "some unique isolated point on the outskirts of geometry," but rather "at the very heart of Euclidean geometrical discourse," namely in "*pure proportion theory*" (TM,

116-17, emphasis in original). Eutocius use of the *epi*-locution is not applied as "spice," as it is in Archimedes, but rather strikes to the heart of the geometrical orientation in Greek mathematical practice. In this sense, Eutocius is much more radical than Archimedes in his promotion of a quasi-algebraic approach, and we must ask what promotes this radicalism.

As we have already seen, Netz insists it is paradoxically the "very lack of ambition, and dependence [on the underlying text]" which makes Eutocius "startlingly original" (TM, 118). Like any good commentator, Eutocius must *make sense* of Archimedes' *epi*-locution "area on [*epi*] line." He does so by isolating it and proving a lemma about the way this locution works in the context of proportion theory. The lemma is not mathematically striking, and in one sense amounts to a simple rephrasing of Greek proportion theory. But from the perspective of *expression*, Eutocius has generalized Archimedes' quasi-algebraic locution and hence the quasi-algebraic tendency latent in Archimedes' original proof. Again Netz resorts to the semantic register of force:

> ... Eutocius' understanding of the situation is strictly governed by Archimedes' terms, so that he needs to make sense of the argument in terms of "area on line." This *forces* Eutocius into originality. (TM, 118, my emphasis)

Eutocius' practice leads him to magnify the anomaly in Archimedes' text, and "this extension, in itself, immediately leads to the algebraization of the text" (TM, 120).

It is only, finally, in the Arabic world that we meet an algebraic *solution* to Archimedes' problem, and here I will focus exclusively on the example of Omar Khayyam. Even in a context where medieval Arabic algebra was a staple, Netz notes that Khayyam's geometric tendencies remain strong: in particular, Khayyam refuses to consider algebraic equations of degree higher than three (third degree equations would correspond to conditions on geometric solids) (TM, 149). However, in some ways "Khayyam downplays geometry." His style of presentation foregrounds cases over geometric properties, and within the study of geometric properties "the foregrounding of equalities over proportions is determined by the overall impulse to provide exhaustive lists. Equalities lend themselves to an exhaustive survey; proportions do not" (TM, 182). Archimedes, on the other hand, "is thoughout motivated by immediate geometrical tasks" (TM, 183). In particular, this contrast dominates Khayyam's solution to the problem of dividing the sphere in a given ratio: it is "not so much a solution to a problem, as a study of the cases arising out of the problem, arranged according to two exhaustive lists of equalities or inequalities" (TM, 167), and the resources of algebra allow Khayyam to handle these inequalities algebraically. Unlike Archimedes, for whom the interplay of the role of the hyperbola and parabola involved in his solution is central, Khayyam is silent concerning a central feature of their interplay (namely the point at which the parabola and the hyperbola are tangents). Netz finds this silence "easy to explain," and he contrasts it strongly with Archimedes' discussion:

> Since Khayyam's study of cases is logically prior to his study of geometrical properties, he is not interested in the geometrical properties of the points that define cases, as long as the points can be stated in terms of his exhaustive lists. For Archimedes, on the other hand, the cases are reached through an investigation of the geometrical properties of the configuration, hence he very naturally states the conditions for the tangencies of the sections. The different priorities determine, quite naturally, which question you pursue and which questions you choose to leave aside. (TM, 168)

This is a good point at which to conclude this brief survey of the trajectory from Archimedes' original solution to Khayyam's algebraic reformulation and solution of the problem. I note only, in summation, that all the key rhetorical terms of Netz's analysis are in place in this passage: the first-order geometric practice of Archimedes versus the second-order deuteronomic, case-oriented structure of Khayyam, the idea of what is "natural" to the respective practices, the idea of practices *determining* priorities, and, finally, the idea of a "choice" within this praxical context of prioritization "determinations."

2.4 Trying to Capture the Uncapturable: The Carnival of Calculation

So far, our focus in the Hellenistic Greek context has been on geometry and the nature of Greek mathematical practice insofar as the quasi-algebraic limits of this geometry are concerned. The central example is drawn from Archimedes' *Second Book on the Sphere and the Cylinder* and concerns, in particular, his "spicy" use of the *epi*-locution. This analysis, even in the broad terms I have presented it, is close-point and may seem picayune, but I believe (along with Netz) that it reveals an important moral. There have been *two* underlying and in fact conflicting assumptions that govern the presentation of the history of Greek mathematics in the work of someone like Heath (to continue to pick on Netz's *bête noire* for a moment). One, which Netz has gone to great lengths to debunk, is that it's OK to use equations liberally to reconstruct what Greek mathematicians were "saying" in modern vocabulary (we find this device throughout the work of Dijksterhuis as well). I take it that Netz's point is *not* that we should bar modern expressions from the consideration of Greek mathematics– to do so entirely would be cumbersome, and in fact Netz does not himself accede to this sort of global stricture. The point is, rather, that when we are trying to understand the way Greek mathematicians *worked*, we must attend to the detail of their textual practice– after all, this is by a far stretch the dominating source we have for reconstructing this practice, as I've already pointed out above.

However, there is a second assumption which I view Netz's work as also extremely useful in debunking, although the extent to which this is a conscious intention on his part is not clear to me. Presentations of Greek mathematics usually center on Greek geometry, and for good reason. But one may come to view Greek mathematics as *nothing but* geometry, and this would be a mistake, indeed. As Netz recognizes, Klein chose to focus on the case of Greek arithmetic, though Netz does not say anything about the consequences of this choice for Klein's work. I personally believe Klein's choice does skew his analysis, but I also think I understand why he makes this choice. Klein is interested in understanding the conceptual roots of Greek mathematics, and the concept of *arithmos* is the most fundamental concept not only for Greek arithmetic, but for Greek geometry as well. It is, effectively, what unifies the treatment of quantity across the panoply of Greek mathematics.

I have often heard it said, usually in philosophical circles, that the Greek preference for geometry involved their strong preference for the definite and their consequent loathing of the indefinite. When Greeks speak of lines, they never mean lines progressing "to infinity," but rather lines with definite, even if unspecified, endpoints. A Greek line is always what *we* would call a *line segment*. Since the numbers are unbounded, proceeding indefinitely, they are suspect from a Greek perspective, so much so that they are even presented in terms of the geometric "pebble diagrams" I have illustrated in the previous chapter.

So the story goes. What Netz's analysis in the third volume of his trilogy, *Ludic Proof*, shows, I believe, is that the story is considerably more complicated, especially (but not exclusively) in the case of Archimedes.[12]

---

[12]Reviel Netz, *Ludic Proof: Greek Mathematics and the Alexandrian Aes-*

The first chapter of the third volume of Netz's trilogy, *Ludic Proof*, is a *tour de force*, both stylistically and in the story it tells, and nothing could be more appropriate given Netz's own emphasis on Archimedean brio. The chapter begins with a report on a piece of detective work: Netz was faced with a single parchment leaf containing the introduction to Archimedes' otherwise lost treatise *Stomachion*. The single extant manuscript page introduces what we would call a tangram problem: how to assemble a diverse collection of triangular and quadrangular pieces into a square. Even the single preserved manuscript page is in bad shape, and Netz began with the hope of improving on the early editor's transcription. He ended with something different and quite impressive: a powerful inroad into Archimedes' style.

In fact, even the fact that the *Stomachion* concerned a trangram exercise was preserved not in the manuscript page itself but in an Arabic fragment and some late testimonies. By adding a single word to the earlier editor Heiberg's transcription, Netz was able to achieve an insight into the origin of the lost manuscript. The word that Netz added was 'multitude', and it occurred in the phrase, "there is not a small *multitude* of figures." This reinforced Netz's conviction that, as he puts it, "the treatise did not foreground geometry. It foregrounded a certain number" (LP, 19). The trangram problem is a problem in plane geometry, to be sure, but not at all like the geometry one would find in Euclid's *Elements*. What Netz's new reading emphasized was not the geometry per se, but rather the abundance, the multitude, of possible solutions to the problem.

Netz himself set about trying to determine all the possible solutions from the description of the tangram in later manuscripts, and was embarrassed that he couldn't come up with the number. He then took the problem to his Stanford colleague Persi Diaconis, a world-renowned expert in combinatorics, and "it took Diaconis

---

thetic (Cambridge: Cambridge, 2009); hereafter cited internally as LP.

a couple of months and collaborative work with three colleagues to come up with the number of solutions – 17,152 – independently found at the same time by Bill Cutler (who relied on a computer analysis of the same problem)" (LP, 20). Although there is no direct confirmation, Netz was convinced that Archimedes resolved the problem himself, and if so it would constitute one of the greatest feats of ancient mathematics. And at the heart of the trangram problem lies a sophisticated interplay between geometry and combinatoric calculation. Here was a showpiece exemplifying the theses of Netz's previous work and then some: that Archimedes chose problems designed to stun the reader with their difficult navigation, and that Archimedes auratic style was driven by the interplay of various "genres" of mathematics: here, specifically geometry and the combinatorics of calculation.

The core of Netz's chapter consists of the analysis of three examples of Archimedes' art (and one from Aristarchus) which stress his "attempt to capture the unbounded." The first comes from his treatise *On the Measurement of the Circle* and involves a fractional approximation to the ratio of the circumference to the diameter of the circle, what we have come to call the value π. Archimedes asserts that this value is less than 3 and 1/7 and greater than 3 and 10/71.[13] In our decimal notation, 3 and 1/7 is approximately 3.142857..., and 3 and 10/71 is approximately 3.140845... The decimal expansion of π is approximately 3.141592... and so does indeed fall between these values. As in the case of the problem from Archimedes *Second Treatise on the Sphere and Cylinder*, here too there are delicate textual isses to be considered, but Netz formulates the upshot of his analysis in five points:

> ... the *Measurement of the Circle* (1) foregrounds the difficulty of calculating the ratio of the circumference to the diameter, it ends up by (5) finding mere boundaries,

---

[13]Netz accidentally inverts the role of 'more' and 'less' in his formulation (LP, 23).

(2) through a process where the interplay of equalities
and inequalities is completely obscured, (3) stressing
the arbitrariness of ending the process at any particular point and of (4) setting out its outcome in any
particular way. (LP, 28)[14]

Together, these five points contribute to the "knockout" style of
Archimedes' presentation, dazzling the reader through an array of
stylistic feints. But at the level of the relation of form to mathematical *content*, Archimedes' solution stresses the thoroughgoing
interplay between geometric and calculational aspects of the problem.

Next, Netz turns to a problem from Aristarchus' *Size and Distances of the Sun and the Moon* to present a "relatively simple example" of the "opaque cognitive texture of calculation" (LP, 37).
For our purposes we may jump directly to the morals Netz draws
from this analysis.

Netz frames these morals in terms of the comparison of cases
where geometry and arithmetical calculation dominate,
respectively:

In the geometrical case, the reading experience is of
a continuous stream of statements whose validity is
parsed and confirmed by the mind through the application of certain known equivalences, of the diagram,
and of the verbal texture: to read is to understand
and agree. In the more arithmetical case, disbelief [sic]
has to be occasionally suspended. The effort of understanding gives way to the effort of merely remembering – even, merely perusing – the complex numerical
values involved. The reader is no longer a *judge*, but
is instead a *spectator*. And with the puncturing of the

---

[14]The numbering, reversed in this summary, corresponds to the previous
more detailed discussion of each of these points.

seamless sequence of parsed deductions, the reading experience is no longer continuous, but is much more discrete, each statement standing on its own. Thus, as the text loses its overarching deductive structure, and as it is no longer under the reader's intellectual control, its overall texture becomes more opaque and difficult. This is the thick texture of calculation. (LP, 40)

Netz's passage beautifully expresses the uncanny experience of the continuous in reading about geometry and the discontinuous experience of reading about the domain of discrete calculation. That is, the reading experience *recapitulates both the stylistic presentation **and** the underlying nature of the mathematical objects involved.* Although Netz's predominant stress is on the *style* of Greek mathematical presentation, the passage shows that the points about style translate readily into points about the experience of *reading* the texts at issue. The whole set of issues surrounding the *reading* of mathematics remain even less attended to than the issue of written style, and so Netz's connection of these two concerns, schematic though it remains, is most welcome.

I will pass quickly over the next pair of Archimedean problems Netz considers, especially since Diksterhuis judges that "the two propositions 9 and 10, as they stand, are about the most indigestible thing in all Greek mathematics" (cited, LP, 41), and jump cut to the end of Netz's chapter, which culminates with the discussion, "A Fascination with Size" (LP, 54). Netz produces a list of 12 large numbers resulting from Greek mathematical calculation, stretching from Hipparchus' calculation of the number of fixed stars (around 1,000) to Eratosthenes' "sieve" method for identifying primes and Aristarchus' heliocentric model for the size of the cosmos, both "approaching infinity" (LP, 58). Right before these two indefinite calculations, we find at the large end of the spectrum two (in principle) definite but enormous numbers supplied in problems set by Archimedes. One is the number Archimedes achieves in his treatise *The Sand Reckoner*, calculating the number of grains of sand needed to fill the cosmos (on the order of 10^{68}), and the second is associated with the so-called problem of the "cattle of Helios," which for its solution would require a number having 206,545 digits (note that by comparison 10^{68} would require "only" 69 digits to write out fully: a one followed by 68 zeroes).

Greek mathematics is distinct among the range of ancient mathematical cultures in its emphasis on geometry. Netz formulates the "standard practice across many cultures" in terms of the representation of "mathematical problems in terms of simple, accessible numerical values" (LP, 47). By way of dramatic contrast, in the "main tradition" of Greek geometry numerals are conspicuously absent: "the only tool used to concretize the problems is the non-numerical, qualitative diagram" (LP, 49). Even more strikingly, perhaps, "whenever numbers assume center stage in Greek mathematics, this is precisely where straightforward presentation fails ... The function of the application of numbers is precisely to thicken the structure of an opaque presentation" (LP, 49). The experience is one of *troubled reading*.

This gives us a new and sharper insight into the old saw that the Greeks were averse to the indefinite or unbounded. The Greeks *were* troubled by the indefinite, but in a rather active sense that is different from avoiding it. We might say that the Greeks *troubled themselves* with the indefinite: they sectored off the domain of the numerical from the "main tradition of Greek geometry" in such a way that when it returned (one thinks here of the Freudian return of the repressed) it came back with the force of a troubled fascination. This is a (and perhaps even the) focal example of the long shadow of the parafinite in a Greek geometrical context.

2.5 Conclusion: Comparing Kleinian Themes and Netzian Narratives

How much does Netz's identification of this long shadow rely on the particular conclusions he derives concerning Greek mathematical style? On the one hand, it is certainly important that Netz pays close attention to the texture of Greek mathematical writing and the way it centers on the role of the lettered diagram. But suppose a devil's advocate, whose counterexamination goes something like this. Netz has taken a weakness of the Greek mathematical style– the geometry-centric incapacity to handle arithmetical calculations naturally– and turned it into a "virtue" of the presentation. The dense, even opaque texture of mathematical calculation– which Netz himself acknowledges– is turned into the virtue of "surprise," even "aura." But anyone who has wrestled with an advanced Archimedean proof *outside the context of the original Greek mathematical culture of its original location* will tell you that these "virtues" of Archimedes' style are in fact nearly insuperable obstacles, and that the reader must effectively rely on the (usually algebraic) glosses of commentators for their assimilation, present author most certainly included. This is not just a matter of our inhabiting a more algebraic mathematical culture, it is simply a matter of our *not* inhabiting the geometric culture of Archimedes. And beyond this, one can even accuse Archimedes, in particular, of being brutal and perverse: laying out traps for his reader, to the extent of even announcing false results as snares! So runs the *advocatus diaboli*.

I won't try to resolve this issue here. The point that I want to make is different, namely, that *either way* we can identify the long shadow of the parafinite in the texture of Archimedean style, whether we value it positively or negatively. To be sure, our "reading" of this shadow will differ on the basis of the morals we draw from Archimedes' style, but the fundamental substance of the issue remains, I think, unchanged. And it bears stressing that in either case, the issue is one which *does* involve the consideration of Greek mathematical style. *In this regard*, I do think Netz has the advantage over Klein, but I don't think it's quite the advantage that Netz thinks he has. Netz doubts, at the beginning of the second book in his trilogy, whether it is possible to express generalizations at the level of "the Greek concept of ..." More probably, he writes, "different Greek thinkers had different views on such issues, as distinct from each other as they are from some modern views. Nor do I think that periods in the history of science are characterized by some fundamental concepts from which the rest follows" (TM, 5-6). I cannot see that Klein commits himself to the latter part of this view, but it is true that he centers his analysis on the role of the concept of *arithmos*, looking in particular at foundational attempts to make sense of the concept. And it is true that *when* Greeks attempted to articulate the concept of a magnitude, they *did so* in terms of the concept of *arithmos*, a definite aggregate of quantity. Within this group, Klein himself finds divergences, surrounding in particular the problems of the existence of ideal *arithmoi* and in accounts of the nature of fractions. No claim is made, so far as I can see, and certainly no claim is necessary for the value of Klein's work, that every mathematician thought *consciously* in these terms, any more than it is necessary for Netz to claim that every Greek mathematician thought *consciously* in terms of the centrality of the lettered diagram.

As I see it, the advantage goes to Netz in other terms, which Netz does himself begin to identify. He says that his aim "is not to argue against Klein's main thesis, but rather to find a historical explanation for an observation that Klein offered mainly on a philosophical basis" (TM, 5). I presume Netz means that Klein's philosophical/conceptual orientation led him to offer a "thesis," which Netz will explain historically. But in fact the problem with Klein's work, to the extent there is a problem, lies not in *Klein*'s basis being philosophical, but in his choosing to focus so heavily on *foundational* issues surrounding the Greek concept of *arithmos*. In particular, this limits Klein's capacity to analyze the complex interplay between the geometric and the arithmetic in Greek mathematics. To be sure, Klein does pick *one* particular case of this complex interplay: the concept of quantity in geometry. But the case is not representative, precisely because it is focused on the foundational question of the status of *arithmos*. That *quantity* is *geometric* is one of the central riddles of Greek mathematics, but it cannot be addressed simply by looking at the foundational investigations of the concept of *arithmos*. Here 'practice' is more fundamental than 'theory' for the almost tautological reason that in this way of distinguishing between practice and theory the former has the wider scope.

On the other hand, if what I have suggested is right, there is no reason in principle for Klein's investigations to be opposed to those Netz provides, and indeed they supply an additional and singular test case for Netz's thesis. This is to say, to put it in Netzian terms, that *foundational* practices are also mathematical *practices*: they cannot be sloughed off simply by calling them "philosophical." How these "peculiar" practices interact with, by reflecting upon, other mathematical practices should be an integral part of an ideal program for investigating Greek mathematical practice. Netz's approach is polemical in steering away from these *topoi*, and the polemic is understandable as a function of Netz's provocative, even provocational emphasis on the fundamental role of practices. But from an ecumenical perspective, at the end of the day, the two approaches are more largely complementary than contradictory.

Netz certainly does win the contest in terms of identifying the "great divide" between geometric and algebraic approaches at an earlier point in the history of mathematics, and in this regard Klein's treatment is demonstrably limited by his historical resources. This is not a fault of Klein's *approach*, simply a fault of his *resources*, and the correction Netz provides is exemplary. But paradoxically, I think Klein, despite his "foundationalist" prejudices, which are real enough, has a more profound sense of *history*. To argue the case is beyond the bounds of what I can accomplish here, but we can at least identify a potential weakness of Netzian narratives. Netz's provocational stance leads him to propose bold, but often very schematic theses. Throughout Netz's trilogy we find the text peppered with claims that "my thesis is simple," as if this were some sort of ultimate virtue. What "my thesis is simple" means, however, is not itself simple, since it involves a semantic range which stretches from 'concise' and 'clear' to 'provocative' and perhaps even strikingly 'paradoxical' in some cases.[15]

Klein's sense of the thetic, of what is *posited* by Greek mathematics, is more sophisticated than Netz's, whose sophistication lies elsewhere. It is for this reason that I have spoken, on the one hand, of Kleinian *themes* and Netzian *narratives*, and, on the other, have emphasized *fractions* and *functions* where Netz emphasizes *problems* and *equations*. Klein's work is indeed more "conceptual" than Netz's, but not in the straightforward, "simple" way that Netz would imply. Klein's emphasis on content, as opposed to Netz's emphasis on form, is best identified at the level of *theme*, and Klein traces out these themes– of the nature of arithmos, of the challenges posed by fractions, of the ideal and the concrete– in carefully modulated variations. Netz tells a different story, one where the overall structure is "simple" but the concrete detail is rich. His modulation, like the calculational in Hellenistic mathematics, is not at the level of theme but at the level of a dense *texture* which is carried along by a simple (but provocative) under-

---

[15]SD, 113: "My theory is simple." SD, 168: "My argument is simple." TM, 2: "At the heart of Unguru's article was a simple claim for historicity." TM, 124: "The argument, quite simply, is developed for only one case . . . " TM, 168: "In fact Khayyam's silence on this point – as well as Archimedes' eloquence – are easy to explain." LP, 230: "The overall structure of the argument is simple."

lying scaffolding of generalizations. Here the devil is in a different sort of detail, more empirical and exemplary, less philosophical and thematic. Kleinian themes and their development resemble more the Platonic dialogues which would come to be Klein's central focus later in his career; Netz's narratives gives us first a Pilgrim's Progress of exemplary instruction in the shaping of Greek geometric practice, and end in a carnivalesque feast of mathematics and literature.

Netz's emphasis on Greek geometric *problems* and medieval Arabic *equations* is attuned to the underlying historical schematism of his narrative: from first-order to second-order treatment of the underlying (geometric) quantities. In the previous chapter, I have focused instead on fractions and functions as what might be called *limiting conceptions* in the ancient and early modern contexts. Klein's discussion of the Greek treatment of fractions is so helpful precisely because it shows what happens to the Greek conception of *arithmos* when it is pushed "to the limit." Netz adopts the more centrist view that Greek mathematics is "about" proportions of whole quantities and leaves fractions largely to the side. As far as equations are concerned, I think it is unequivocal that we find a full-blown treatment of them only in the medieval Arabic context, but one where the underlying quantities remain geometric in nature (as Netz recognizes). It is *only* in the early modern context that we accede to the full-blown mathematical development of the concept of *function*, which serves as a "limit" of the underlying idea of functional *relation*, already in evidence even in the Hellenistic Greek context. These "limiting" cases are helpful as test cases precisely because they *are* limiting cases, and so allow us to gauge the extent of "mathematical abstraction" along the trajectory of its historical development. Such a way of speaking, personifying the concept of abstraction as it does, would no doubt be anathema to Netz, given his ideological insistence on mathematical practices, and yet I think that once the *ideologies* of conceptualism and praxism are suspended, the fruits of the two perspectives are much more reinforcing than either ideology would be prepared to admit.

And the long shadow of the parafinite? Our preliminary encounter suggests that this shadow is cast by an embedded inversion of the "standard" figure-ground relation. That is, in the standard figure-ground relation we find a definite figure against an indefinite background, what we might call a "horizon." For example, a lettered geometric diagram is located in an ambient "space," indefinite in extent, which serves as its background. But now in a different sense, we find an explosion of the uncapturable located relative to a definite "background" geometric configuration (think, for example, about Archimedes' approximation of the value of π). To be sure, the "background" against which this explosion of calculation occurs itself has an indefinite background in turn, and that is precisely why I have referred above to an *embedded* inversion of the figure-ground relation. We do not witness the power of the indefinite directly, but within a definitely circumscribed context where it enters as brilliant, dazzling, obfuscating, carnivalesque, auratic, even contaminative. Because of the ambient context of definition, we are not seeing the parafinite directly, only the long shadow it casts.

That is a first identification of a dramatic crux, but one we are still in the process of tracking.

Scene Two

Early Modern Mathematics And Physics: Galileo And Leibniz

Chapter 3

Galileo and the Birth of the Parafinite

3.1 Introduction

Taking Alexandre Koyré's work *From the Closed World to the Infinite Universe* as a point of departure, scholars have often assumed that it is the status of the *infinite* which is decisive for modern metaphysical attitudes and, specifically, what distinguishes them from earlier approaches to metaphysics.[1] (Here I am using the term 'metaphysics' generically to refer to the treatment within the European tradition of fundamental philosophical investigations concerning the nature of reality.) On this view, largeness as such is not fundamental, and it is this view which I am contesting along the line of this volume. In this particular chapter I focus on a brief episode in which Galileo posits a specific concept of the parafinite as a quantity existing between the finite and the infinite. It is true that Galileo's treatment of the parafinite was rapidly assimilated to the ongoing development of the mathematical infinite,

---

[1]Alexandre Koyré, *From the Closed World to the Infinite Universe* (Baltimore: Johns Hopkins University Press, 1957).

but what Galileo emphasizes is the problem of *large*, not infinite, numbers, and it would be a mistake not to insist on this revolutionary turn in his thinking. What is most important is the radical novelty of the venture Galileo initiates by proposing a parafinite quantity. It may well be the case that the infinite, as Koyré and others have suggested, became a standard against which the status of modern metaphysical enterprises was measured. If this is so, then modern metaphysicians will evaluate their own enterprises in terms of the way they handle the distinction between the finite and the infinite. But this is not my concern here: instead, I am interested in what *drives* these enterprises, independently of how the people who conduct philosophical investigations *frame* them. And in terms of these deeper motivational roots, it is the *problem* of the status of the parafinite which leads to the deeper characterization of modern metaphysics specifically and modern philosophy more generally. Galileo's radical suggestion of a parafinite quantity bears on the crucial question of the systematic use of numbers to describe reality, but sadly this was quickly effaced by its assimilation into the early modern treatment of the infinite. Yet traces of the Galilean parafinite linger on long after the notion of a quantity midway between the finite and the infinite is explicitly abandoned.

What makes a number large? Are large numbers on a par with small numbers? If we knew how sizeable largeness is, could we make do with what seem to be large enough numbers? Or do we need to make recourse to infinite numbers? Are infinite numbers properly understood as *large* numbers? (*How* large? Incomparably large?) Above all, what makes anything large? Galileo's category of parafinite quantity provides a first attempt to answer these questions *directly*.

There are many questions about the parafinite that Galileo's treatment will not yet resolve. Perhaps the determining issue is this: unless we are dealing with infinite numbers (and perhaps even then), there just *aren't* intrinsically large numbers: it is a mistake to think that there are some numbers which *are* large. What counts as large and what counts as small will depend on context and the fixing of a particular scale of measurement. We have already seen the passage from Plato's *Republic* in which numbers are treated as "summoners," that differences of greater and smaller among them point "beyond themselves" and cannot be understood in terms of the numbers themselves, intrinsically. Such summoners "lead us towards truth"[2] rather than *being* truth itself. But what is this "dependent" status of the greater and smaller? To suggest that sufficiently large numbers (whatever they might turn out to be) are parafinite may seem to confuse our limited capacity to imagine quantities with the non-psychological nature of the notion of finitude.

Galileo does not address these concerns explicitly, and it will take time for them to come into focus. The problem, ultimately, is not that an account of largeness requires an appeal to context, but rather that context continues to be thought of as something which is "extra-added" to numbers. Galileo does make a first step in tying the notion of context directly to number in suggesting a parafinite quantity, and seeing context as an intrinsic part of our conception of number just is a core issue in addressing the problem of the parafinite.

For the moment, our focus must rest on the fact that the notion of the parafinite emerges naturally in the context of Galileo's treatment of problems concerning the strength of materials in the First Day of his *Two New Sciences*. This confers a historical support for the significance of those issues which, in some sense, Galileo was the first to broach so explicitly, and which are even more critical today. His discovery provides a natural starting point for consideration of larger questions about the status of the parafinite in the modern age.

---

[2]Plato, *Republic*, 525a, Grube/Reeve translation, 197.

3.2 On the First Day Galileo Created the Parafinite

From the beginning of Galileo's *Two New Sciences*, the issue which organizes the conversation is that of *scale*. A scale is a "yardstick" against which we measures things, but it will only be a useful scale if the objects we must measure are close enough to this scale. For example, we can measure something in the range of fractions of inches using a yardstick, and by laying the yardstick end-on-end we can even measure something a number of yards in size. But it would be pointless to try to measure something with a yardstick which is too small: for that we would perhaps use a micrometer. Similarly, if the object is large enough (say the distance from New York City to Los Angeles) it wouldn't make any sense to measure it with a yardstick either. Objects plus an appropriate context pick out the appropriate scale for measurement.

Galileo is concerned with how the largeness of a body is related to its strength. Does a larger mass break under its own weight more readily than a smaller one? Apparently so, but how is this to be related to the size of the object? This means, actually, that *two* scales will be involved: a scale for measuring the largeness of bodies, and a scale for measuring the strength of these same bodies. Then these magnitudes must be compared, which involves a *third* scale appropriate to these "size versus strength" comparisons. Intuitively, we can already see that this notion of scale massively extends the classical notion of the dimension of a quantity as linear, planar or volumetric. The problem of relating quantities associated with different scales constitutes a massive extension of the Hellenistic and medieval Arabic attempts to bring, say, linear and planar quantities into relation.

Galileo's *Discourses* are organized in dialogue form, and the character Salviati serves the role of Galileo's spokesman. Salviati makes a first step in answering questions about size versus strength by distinguishing between perfection and strength. He recognizes that there are examples of machines more perfect at the large than at the small scale, but so far as *strength* of materials is concerned the important point is that there are limiting size constraints:

> the mere fact that it is material makes the larger framework, fabricated from the same material and in the same proportions as the smaller, correspond in every way to it except in strength and resistance against violent shocks [*invasioni*]; and the larger the structure is, the weaker in proportion it will be.[3] (GO, 51)

Salviati tells us that *perfection* is more closely obtainable when the machine is large, which is not surprising given that a larger machine is capable of proportionately more accurate tooling. (It is a lot easier to make an accurate yardstick than a comparably accurate micrometer; notice, however, that in such a comparison we are judging each instrument relative to its *own* scale rather than one common to both, and *only then* using a common scale to compare the difficulty of fashioning them.) But, on the other hand, *strength* becomes more of a problem as the machine grows. Simply put, it must bear its own weight.

Bodies are often, though not always, broken by a violent rather than a gentle motion, a smashing rather than a smushing. So Salviati insists that in an analysis of the strength of materials it is resistance to *violent* treatment which is most important. But

---

[3]Galileo Galilei, *Two New Sciences*, trans. Stillman Drake (Madison: University of Wisconsin Press, 1974) Galileo Galilei, *Opere di Galileo Galilei* (Florence: Edizione Nazionale, 1890-1910), vol. 8, 51. All translations will be taken from Drake's translation, but page references will be given to the pagination in the *Opere*, vol. 8, which is given in the margin of both Drake's and Crew's translations (Galileo Galilei, *Dialogues Concerning Two New Sciences*, trans. Henry Crew (New York: MacMillan, 1914, repr. Dover and also Northwestern University Press)). All references to Galileo will be to the 8th volume of the *Opere*, and will be given in the text as GO.

when a motion is violent there can also be a violent disruption of the surrounding matter, and this leads to the question whether a violent motion can actually *empty* a portion of space of all matter for some length of time, creating a vacuum or void. Sagredo, one of Salviati's conversation partners, says a bit later that we must say that by force [*violenza*] (or contrary to nature) a void is sometimes to be admitted (GO, 60). In traditional Aristotelian physics such a void was deemed impossible, and so a void, even only under special violent conditions, constitutes a departure from tradition which both Salviati and Sagredo share. Galileo places this remark in the mouth of Sagredo, who goes on to declare that "in my opinion nothing is contrary to nature save the impossible, and that never happens" (GO, 60). As Stillman Drake puts it in the introduction to his translation of the *Two New Sciences*, Sagredo

> speaks for Galileo at the middle stage of his development—from his move to Padua in 1592 to the essential completion of his mathematical physics a quarter-century later... We find him raising questions that had puzzled Galileo earlier, and taking positions that Galileo had once considered and then rejected. Indeed, some of the questions Sagredo raised still puzzled Galileo, as is evidenced by Salviati's replies to them.[4]

When Sagredo and Salviati converse, Galileo stages a dialogue between two stages of his own career. Galileo accepted the existence of vacua, or voids, both in this middle stage and at the later stage during which the *Two New Sciences* was written, so it is not on this issue that Sagredo and Salviati differ. What, then, is the difference between them, and how has Galileo's position changed? Salviati's position blends a subtle appreciation of the *unnaturalness* of voids, following Aristotle, and the potential for their introduction by *violent*, unnatural motions. But although he appeals to them, at this

---

[4]Drake, Introduction to *Two New Sciences*, xiii.

point in the *Two New Sciences* he defers an explanation of how these voids are created.[5]

Voids are important for Galileo because they create gaps, and he will use these gaps to explain how things fall apart. On the other hand, Galileo will use the tendency which bodies have to close gaps (to avoid voids!) as an explanation for what makes them hold together. In the service of this type of explanation, Galileo introduces a distinction between finite gaps and gaps which are smaller than finite without being nothing at all. Galileo is at pains to hedge his commitments regarding finite, perceptible voids, in part because these were the kind explicitly denied by the Aristotelians. In one place, he puts thoughts about them into the speech of Sagredo and in another case he simply defers discussion. But Galileo will use *some* appeal to voids to explain how bodies stick together: the basic idea is that when there are *little* voids between pieces of matter the matter strives to close these gaps. This cohesion will be strong enough to hold the body together *except* in the case of violent motions: in this way Galileo can explain both why bodies usually hang together and why they sometimes fall apart.

Although Aristotle distinguished between natural, gentle motion and unnatural, violent motion, Galileo is using the distinction in a very different way. For Galileo the distinction between natural and violent is a distinction in *scale*–which takes the emphasis off the distinction between the natural and unnatural, focal for Aristotle, and shifts it to the distinction between the small and large. Galileo begins to argue for this different perspective by appealing to common examples:

> For who does not see that a horse falling from a height
> of three or four braccia will break its bones, while a dog
> falling from the same height, or a cat from eight or ten,

---

[5]see GO, 112.

> or even more, will suffer no harm? Thus a cricket might
> fall without damage from a tower, or an ant from the
> moon. (GO, 52)

He then goes on to indicate that the issue is that strength is not
in constant proportion to size: as the size increases the weakness
of the material increases even faster. "For it can be demonstrated
geometrically that the larger ones are always proportionately less
resistant than the smaller" (GO, 51).

Since strength of materials is a function of how well bodies stick
together, Galileo must first provide some account of the nature
of bodily coherence in order to explain the circumstances under
which materials will fracture or break. The key axiom organizing
his position is that "whatever may be the tenacity and mutual
coherence of the parts of this solid, provided only that it is not
infinite[ly strong], it can be overcome by the force of the pulling
weight C, of which the heaviness [*gravità*] can be increased as much
as we please..." (GO, 55). That is, tenacity and coherence are
finite quantities, at least in everyday materials like wood. But this
overcoming of the coherence of a body only occurs at a *given* finite
value: so long as this value is not attained the body will continue
to stick together. Fracture of a particular body, then, is an issue
which breaks the symmetry of finite numbers by *fixing* a scale:
forces below a certain value are small in the sense that they do not
fracture the body whereas forces above the same value are large in
the sense that they do.

Physicists speak of symmetry breaking when a balance in the
physical world is broken by some (generally small) variation in otherwise balanced conditions. For example, if a coin is precariously
balanced on its edge, a slight breeze could topple it in one direction or the other. In the case of materials, bonding forces are in
a kind of competition with forces of strain. As long as the strain
is not too great, the material body will stick together. But as
the strain increases, a critical threshold is reached where bonding

forces and straining forces match each other. At this point, if the straining forces are increased even slightly, the body will suddenly fracture. (Actually, this is a bit of a simplification, since fracturing often happens over a long period of time, but for our purposes, the simplification will do.) The critical value where strain overcomes bonding establishes a *scale* against which strains can be measured. Below the threshold value, no fracture; above it, fracture will occur.

In Galileo's discussion, a scale distinguishing small from large effectively replaces the Aristotelian distinction between the natural and the violent. But how is the existence of this scale to be explained? The appeal to scale explains *when* bodies will, and will not, cohere, but in order to account for this scale Galileo must further explain *how* they stick together. That is, he must explain the *nature* of the bonding which holds them together. *After* providing concrete examples to illustrate some basic features of coherence, Salviati goes on to tell his interlocutors that he sees coherence deriving from two sources: negatively from *horror vacui*, or the desire nature has to avoid a vacuum, and positively from a gluey or viscous substance which binds firmly together the component parts of the body (GO, 61). Salviati argues that a substance which is held together solely by virtue of the avoidance of a vacuum is the most weakly bonded, and that by virtue of its weak bonding this substance will be continuous and fluid. This is the most basic form of coherence, and Salviati gives water as an example:

> . . . it is reasonable to argue that the minimum [particles] into which water seems to be resolved, since it has less consistency than the finest powder (or rather, has no consistency at all), are quite different from quantified and divisible minimum [particles], and I cannot find any other difference here besides that of their being indivisible. (GO, 86)

By implication, the *other* force of cohesion, the gluey or viscous type, will be that which binds things as *finite*, divisible particles.

This positive source of bonding is a second level of bonding added on top of the previous, negative one. This physical explanation on two *separate* levels requires Galileo to introduce a distinction between finite voids– avoided on the lowest level of cohesion– and indivisible voids, which explain the gluey viscosity at the higher level of coherence.[6] This will be the conceptual division of labor that leads Galileo to appeal to the parafinite. Galileo uses the notion of vacuum cohesion to explain, for example, why water cannot be pumped (i.e. sucked) up beyond a certain height: at this height the force of vacuum fractures the column. But this only allows Galileo to explain the weakest sort of coherence, that of a non-viscous fluid. Salviati will go on to suggest, however, that there are *non-finite, indivisible vacua* and that we use these to explain that *additional* cohesion which extends over and above the cohesive force of the vacuum (GO, 66). The force of the vacuum is the result of nature's tendency to avoid finite vacua; the indivisible vacua will account for the further gluey viscosity of more coherent substances. In this way, the physical problem of cohesion prepares the way for the basic role played by Galileo's distinction between the finite and the infinite, and ultimately the parafinite.

On the face of it, the simplest way to understand the situation might seem to be to distinguish between finite voids– the kind avoided by nature, and infinitely small ones, which nature allows. However, the fact that Galileo does not admit any strictly infinite quantities (and therefore also no infinitely small ones),[7] compli-

---

[6]This may seem backwards at first, but think of it this way: it is more fundamental to avoid finite voids, which are bigger, than to avoid the smaller, indivisible ones.

[7]Here I follow Eberhard Knobloch; see his article, "Galileo and Leibniz' Different Approaches to Infinity," *Archive for the History of Exact Sciences* **54** (1999), 87-99, here 94. Although there are some delicate (but ultimately, I think, crucial) points on which Knobloch and I differ, I am in broad agreement with his interpretation of Galileo's treatment of indivisibles, and so for the purposes of this paper I will largely adopt Knobloch's approach. By contrast,

cates matters. He attempts to deal with this problem by speaking of the non-finite, indivisible vacua as non-quantitative, and it will take some work to see what this might mean. Earlier in the dialogue Sagredo had already illustrated how small finite quantities could sum to an extraordinarily large quantity if sufficiently many of them are heaped together, noting that in an army "every individual soldier was paid with pennies and farthings collected by general levies, although a million in gold was not enough to pay the whole army" (GO, 66). Galileo's suggestion is much more conceptually ambitious, however, since it now involves a summation of infinitely many non-quantitative indivisibles. But if there are not infinitely large quantities, how can we add together infinitely many indivisibles? This cannot mean that we sum an infinite *number* of indivisibles, since there are no strictly infinite quantities. What happens if we attempt to sum an infinitude of forces? It seems clear to Salviati that if we do this we arrive at an infinite quantity.[8] But then how can we sum infinitely many indivisible voids

---

the discussion of Galileo's treatment of indivisibles in Maurice Clavelin's "Le Problème du Continu et les Paradoxes de l'Infini chez Galilée," *Thales* **10** (1959) 1-26, is seriously deficient along lines that Knobloch's essay makes clear. Nonetheless, I remain broadly sympathetic to Clavelin's thesis that Galileo's approach to the continuum breaks decisively with his scholastic predecessors as a direct consequence of the dynamic demands of Galileo's treatment of uniform acceleration; here see also Maurice Clavelin, *The Natural Philosophy of Galileo*, tr. A. J. Pomerans (Cambridge: MIT Press, 1974), Part III, The Birth of Classical Mechanics, especially Chapter 6, The Geometrization of the Motion of Heavy Bodies (Part I), 277-323. For an opposing view, see A. Mark Smith, "Galileo's Theory of Indivisibles: Revolution or Compromise?," *Journal of the History of Ideas* **37** (1976), 571-88. For an overview of Galileo's commitment to atomism, see William R. Shea, "Galileo's Atomic Hypothesis," *Ambix* **17** (1970), 13-27. None of these authors addresses Galileo's assertion of a quantity between the finite and the infinite.

[8]From a contemporary perspective, this will of course look like a fallacy, but I would caution the reader against making the obvious inference that Galileo has committed a simple mistake. I will return to this issue below, and again in detail in the next chapter; I also discuss it in "Leibniz on the Indefinite

within a finite quantity?

In order to illustrate the notion of indivisible voids, Galileo appeals to an analogy which I will describe in a general manner here and in more detail in Appendix B.

Think of the example of a wheel rolling without slippage on a surface. As the wheel makes one full revolution, it traces out a distance along the surface equal in length to the entire circumference of the circle. Look at the bottom part of the following figure, which is taken from Galileo's own discussion (I'll discuss the top part of the figure briefly below and in greater detail in Appendix B):

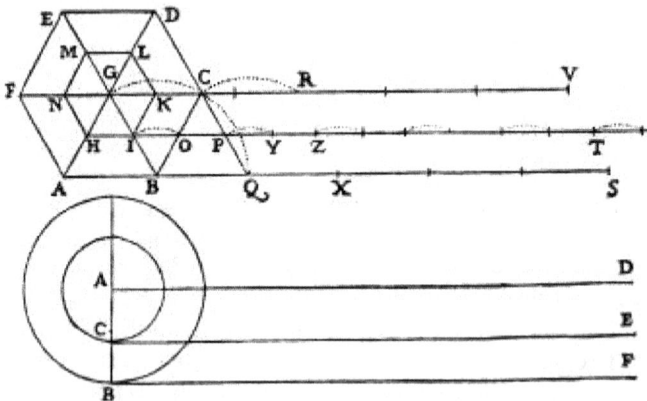

Figure 3.1: Galileo's rolling hexagons and circles[9]

In this picture, the outer circle originally touches the surface at the point B, and as it runs through one full revolution this circle traces out a path from B to F. But at the same time, the smaller circle that runs through C traces out the path from C to E, which is the *same* length as the path from B to F. Although the circle

as Infinite," *The Review of Metaphysics* **51** (June 1998), 849-871, especially 859-865.

[9]Figure taken from the Northwestern edition of the Crews and de Salvio translation, 21.

based at C is much smaller than the circle based at B, and so has a shorter circumference, it traces out a path just as long as the path as the one traced by the larger circle. How is that possible?!

Galileo's answer is to say that in the case of the smaller circle tracing out a length much greater than its own circumference, little indivisible voids are interposed all along the path which the circle traces out. These voids expand the path so that it can be larger than the circumference of the smaller circle and make it just as long as the path traced out by the circumference of the larger circle. For even smaller circles still, which must also trace out the same length, even more indivisible voids will have to be added.

Galileo is working in a tradition, inherited from Aristotle, that distinguishes between continuous and discrete magnitudes; the goal of an account of the mathematical continuum, as given here, is to give an account of the former, continuous magnitudes. Galileo's presentation in the *Two New Sciences* was no doubt influenced by the work of his disciple, Bonaventura Cavalieri, which was presented in his monumental *Geometry Advanced in a Way by the Indivisibles of the Continua* (over 500 pages!) in 1635, three years before the appearance of the *Two New Sciences*. Both Cavalieri's debt to Galileo and the extent to which Galileo was responding to Cavalieri in the *Two New Sciences* are subject to debate, but what is generally agreed is that the debates about the status of indivisibles dominated the minds of Galileo's followers in the years following the publication of the *Two New Sciences*, much as debates about the status of infinitesimals were to dominate the French Academy of Sciences following Leibniz's writing on the foundations of the calculus.[10]

---

[10]For a nice discussion of these and related issues, see Michael Segre, *In the Wake of Galileo* with a forward by I. Bernard Cohen (New Brunswick: Rutgers University Press, 1991), 56-60 and 69-78. On the basis of evidence from the correspondence of Cavalieri with Galileo, Segre even mentions the possibility that Galileo may possibly have been at work on a project on indivisibles during the 1620's; see 71.

What is most important about Galileo's example is that it offers a picture for how an *infinitude* of indivisible voids can be located in a *finite* magnitude. The issues which this picture evokes continued to be of intense concern among Galileo's followers, and in fact Cavalieri's method of indivisibles, which need not detain us here, was criticized by the mathematician Guldin on the ground that an infinite quantity of indivisibles cannot be said to make up a finite magnitude.[11] In Galileo's dialogue the problem is connected very directly to the physical problem of coherence, and this may have given him a heuristic advantage for dealing with it over Cavalieri, who set the physical concerns of his mentor off to the side. Although Cavalieri spoke of the motion of lines, he made no significant attempt to tie this to the physics of bodies and their coherence, as Galileo did.

Galileo resolves the problem of the stretching of the continuum by focusing on the rolling motion of the circle along the line. If we think of the track which any given circle traces out in such a motion, we may think of this as a resolution of the circular figure into the linear trajectory traced out by its point of contact. Going one step further, we could even think of unwrapping the circle along the line by cutting the circle, say at its base point B and then flattening it out along the length of the line BF. In so doing, we break the curvature of the circle at every single point along its circumference, and so, as Galileo puts it, resolve the continuum down into indivisibles at one fell swoop.

This is, in a sense, completely different from what would happen it we did the analogous thing with a hexagon $ABCDEF$ (see the upper part of Figure 3.1). Unwrapping it along the line AS, we would only break the hexagon at the points A, B, C, D, E, and F. In this case, instead of resolving the continuum down into indivisibles, we would simply break it into six pieces, each of which is itself a continuum of finite length. (I will discuss this hexagonal

---

[11]See Segre, op. cit., 76.

case more extensively in Appendix B.)

This registers an *absolute distinction* between the finite and infinite level, and so Galileo must go on to give some account of the distinction between the finite and the infinite. From the above example, notice that it wouldn't matter how many sides the polygon in question had: we looked at the example of a six-sided polygon, or hexagon, but the same point could have been made for a polygon with 1,000 or even 1,000,000 sides. So, in particular, Galileo must give us an account of the distinction between the finite and the infinite which distinguishes the infinite from the finite and large. In fact, he will ultimately do much more than this, interposing a notion of quantity *between* the finite and the infinite, thus becoming the first person explicitly to recognize a notion of the *parafinite*!

Galileo takes it that in fact the greatly finite, i.e. the finite and large, is at *furthest* remove from the infinite; indeed, he will eventually take the infinite to be unity. He begins to motivate this position by stressing the *contrast* between the great and the infinite. He notes, for example, that as we move to larger and larger numbers the square numbers become sparser and sparser. That is, although the numbers proceed by unit steps, the square numbers spread further and further apart from each other. The difference between the square of 2, which is 4, and the square of 3, which is 9, is only 5, but the difference between the square of 11, 121, and the square of 12, 144, is already 23; as we proceed to larger and larger numbers these differences will only become larger. But as Galileo notices, in the infinite there are as many squares as roots, since each number can be paired off with its own square. Thus, he concludes that "to go to ever and ever larger numbers is to move away from infinite number" (GO, 79). Since the gaps between successive squares become larger and larger, the square numbers become sparser and sparser: between 1 and 100 there are 10 square numbers, between 101 and 200 there are already only 4.

Several pages later infinity is explicitly associated with indivisibility, and finitude with the process of successive division of the continuum: through finite division of the continuum one "will never achieve the division of the line into all its quantified parts, even if he goes on forever; and as to its indivisibles, he would be so far from arriving at the desired end by that path that instead, he would be traveling away from it" (GO, 82). In order to support his contention that finite division becomes less and less like infinite division as more divisions are made, Galileo also appeals to a physical example: he notes that the more finely a crystal is ground, the less transparent it becomes, but water is extremely transparent, which suggests that in this latter case the division is infinite (GO, 86). This is a particularly fine illustration of how Galileo uses mathematical arguments and physical examples to reinforce each other.

What Galileo needs now is an account of the distinction between a finite and an infinite division of the continuum, and it is this which will force him to the notion of a parafinite quantity. In the previous sentence, the expressions 'finite division of the continuum' and 'infinite division of a continuum' are ambiguous, but deliberately so: we may be speaking of the division of the continuum into finitely many parts or into parts of finite size, and in the latter case of a division into infinitely many parts or into non-quantitative indivisibles.[12] For Galileo, there is (at least) a

---

[12]Here I depart from Knobloch, who recognizes that the notions 'finite/infinite' do not distinguish between numerative and quantitative aspects, but thinks that all ambiguity is resolved by looking at the way that Galileo distinguishes between quanta and non-quanta (see Knobloch, "Leibniz and Galileo," 90-1). My point of view, on the other hand, is that the ambiguity is intentional but harmless precisely because Galileo admits only finite aggregates of finite parts and parafinite aggregates of parafinitesimal parts (as will become apparent below, it only makes sense to speak of parafinite *collections* of parts as including parafinitesimals; see the conclusion of the current chapter). Of course, to see how Galileo is able to accomplish this requires addressing his distinction between parafinite and infinite division, which Knobloch does not do.

problem with the division of a limited continuum (i.e. one finite in length) into an infinite number of *finite* parts: it is clear to him "that quantified parts actually contained in their whole, if they are infinitely many, make it of infinite magnitude" (GO, 80). Salviati seems to recognize this as a problem only in the case of a continuum of finite length. But the division of such a continuum into a finite number of parts cannot be exhaustive unless we admit ultimate (and hence indivisible) parts of finite magnitude. *So Salviati commits to the existence of the parafinite between the finite and the infinite*:

> Speaking of discrete quantity, it appears to me that between the finite and the infinite there is a third, or middle, term; it is that of answering to every [*ogni*] designated number. Thus in the present case, if asked whether the quantified parts in the continuum are finite or infinitely many, the most suitable reply is to say neither finite nor infinitely many, but so many as to correspond to every specified number. (GO, 81)

Galileo understands this in terms of an unspecified quantity, so that we can take the line to have indefinitely many (quantified) parts, i.e. *however many parts we please*. As Galileo puts it:

> To do that, it is necessary that these be not included within a limited number, because then they would not answer to a greater [number]; yet it is not necessary that they be infinitely many, since no specified number is infinite. And thus at the choice of the questioner we may cut a given line into a hundred quantified parts, into a thousand, and into a hundred thousand, according to whatever number he likes, but not into infinitely many [quantified parts].[13] (GO, 81)

---

[13] Compare Leibniz's defense of the validity of terminological appeals to the

Galileo's commitment to a parafinite quantity is an explicit reaction against the Aristotelian distinction between the actual and the potential. Aristotle had explained the distinction between the finite and the infinite in the context of the divisibility of the continuum by saying that although at any given point the continuum will only have been *actually* finitely divided, it is *potentially* capable of infinite division, which is to say that we may go on dividing it *indefinitely*. But Galileo rejects this way of distinguishing between the finite and the infinite, insisting instead that the finite, the parafinite, and the infinite can all be taken as potential or actual as you choose (GO, 81). Indeed, Galileo's introduction of the parafinite is explicitly posed as an *alternative* to this Aristotelian distinction, and it works analogously without having to distinguish between the actual or potential character of a division. Just as Galileo retained some of the power of Aristotle's denial of physical voids while nonetheless enlisting them, so here on the mathematical front he revises Aristotle's approach to quantity in a way which achieves what Aristotle did by distinguishing between potential and actual, yet without requiring this distinction. In both cases Galileo's departure from Aristotle is subtle and maintains much of Aristotle's power while casting it in a very different light.

Galileo's innovative introduction of the parafinite marks a kind of natural midpoint to the dialogue of the First Day, since we began with very concrete problems about the strength of materials before wandering off into speculations about the nature of the finite and the infinite. Galileo's description of an infinite division of the continuum into indivisibles at one fell swoop comes at this point (GO, 93). As we saw above, if we take a polygon and roll it along a surface through one complete revolution, it will make a

---

mathematical infinite by appeal to finite quantities as large or small as is necessary [*aussi grand et aussi petit qu'il faut*] in the August 1701 letter to Pinson, G. W. Leibniz, *Mathematische Schriften* (Hildesheim: Olms reprint, 1971), IV, 96. I discuss this point in the context of Leibniz's letter to Varignon in the next chapter.

mark on the surface each time a side lands (i.e. at the points B, Q, X, etc. in the top half of Figure 3.1). By analogy, when we roll the circle along the surface, each point on the circumference of the circle hits a point on the surface as the circle rolls along. In rolling the circle along the surface we have therefore marked *every* point on the surface, and so resolved it into indivisible points in one fell swoop. As I suggested above, we could think of this resolution in terms of breaking the curvature of the circle as it is laid out along the line.

After this example, the conversation will increasingly turn back in the direction of practical problems, with the intent being to show that the abstruse speculations about quantity are in fact integrally related, not only to physical problems, but also to how we should *reform* our perspective on physical problems. Galileo is stressing that the theoretical speculations lead directly back to physical pay-off: they provide new conceptual tools for handling physical problems. Galileo goes on to provide a sophisticated account of the contributions of finite and indivisible vacua to the phenomena of condensation and rarefaction and then proceeds to consider further examples which move in the direction of accounting for the characteristic values which determine various material properties such as the point of fracture. This is accomplished through the elaboration of a notion of characteristic frequency (GO, 141). Galileo's trajectory points in the direction of canonical modern problems already exemplified in Galileo's preoccupation with vibrating strings, thus tying his program back to the earliest Pythagorean investigations. The work surrounding these problems has continued to be motivated by the example of a string vibrating and producing a tone. Appropriately, the mathematics used to describe this physical phenomenon is called harmonic analysis.

3.3 Conclusion: Parafinite Lost

The significance of Galileo's discussion for the developing establishment of modern scientific enterprises has been and remains clear, but what bearing do these points have on the issue of the status of modern *meta*physical enterprises? In order to see this it is necessary to investigate further the terms in which Galileo couches his description of the infinite in the First Day of the *Two New Sciences*. Galileo's introduction of the parafinite effectively replaces the Aristotelian distinction between the potential and the actual. This is so, in particular, because the idea of a parafinite division of the continuum replaces the notion of the continuum as potentially infinitely divisible but only actually finitely divided at any given time.[14] But by insisting that the *parafinite* in particular may be taken as actual or potential as you choose, Galileo suggests that we may speak of an *actual* division of the continuum into a *parafinitude* of parts. What is Galileo's motivation for this novel way of proceeding?

A first response to this question must be that, with the strong appeal to an actual conception of infinity in the metaphysics of the heretical Giordano Bruno, the metaphysical infinite had acquired a political and theological charge that threatened to disrupt any appeal to it. The aftershocks of Bruno's burning at the stake in 1600 can be observed in the many qualifications with which Galileo traps out his discussions of the infinite in the First Day of the *Two New Sciences*. At one point Salviati prefaces some speculations involving the nature of the infinite with the remark that they will involve "a fantastic idea of mine which, if it concludes nothing necessarily, will at least by its novelty occasion some wonder," but then goes on to offer to withhold them since "perhaps it will seem to you inopportune to digress at length from the road that we started on, and hence will be distasteful" (GO, 73). At another point Salviati

---

[14]Aristotle, *Physics* 206b12.

cautions concerning his account of condensation and rarefaction of bodies: "If anything in it pleases you, make capital of that; if not, ignore this as idle, and my reasoning along with it, and go search for some explanation that will bring you more peace of mind. I repeat only this: we are among infinites and indivisibles" (GO, 96). At a third point he anticipates the concerns of his interlocutors by suggesting that if they do not like his explanation they should find something better themselves (GO, 105)– something we've all, no doubt, wanted to say at some point or another, and maybe even have!

Yet the appeal to the parafinite cannot be *explained* in these terms, or in any others like them, for the alternative Aristotelian approach, distinguishing between an actual and a potential infinity, was also still available. Why did Galileo insist that the parafinite, in particular, be understood as either actual *or* potential, as you choose? We are dealing, that is to say, with a quantity which is both as large as you choose and actual or potential as you choose.

Maurice Clavelin has argued that Galileo's rejection of the Aristotelian appeal to a potentially infinite continuum which is actually only finitely divided is, in fact, consonant with Aristotle's argument that whatever exists in actuality must first have existed potentially; and there is nothing to prevent what exists potentially from becoming actual in due course, so that there is no justification for extending this privilege to certain properties but not to others.[15] This might have as a consequence that a commitment to the potential infinite already brings a commitment to the actual infinite in its wake. If this were to be the case, the advantages of appeal to the potential rather than the actual infinite are largely or entirely squandered. It is debatable whether Clavelin has correctly understood the force and implications of Aristotle's argument, or whether a commitment to the actual infinite effectively follows from

---

[15]Maurice Clavelin, *The Natural Philosophy of Galileo*, 315; Clavelin refers to the passage in Aristotle's *Physics* at 207b 17-18.

the claim. Certainly Aristotle himself never drew this conclusion with respect to the infinite, but an argument *was* made connecting the actual to the potential infinite in this way by early modern figures such as Pascal, among others.[16]

Galileo avoided any such concern by focusing on a conception of quantity, whether finite or parafinite, which is independent of the distinction between the actual and the potential. Instead, Galileo simply argued that an infinity of finite parts could not be contained in a (finite) continuum, and neither could the number of finite parts in the continuum be finite, for this would imply an ultimate division of the continuum into indivisible finite, i.e. extended, quantities. The recourse to the parafinite follows as an immediate consequence of this situation. The now canonical alternative, which I discuss below, is to view Galileo's assertion that an infinity of finite parts cannot be contained in a finite continuum as an error.

Ultimately, then, Galileo needed an *indeterminate notion of quantity*, which is to say: one that would not be tied to traditional metaphysical distinctions between either the finite and the infinite *or* the actual and the potential. What was involved with and implied by such an indeterminate notion of quantity was not something Galileo was able to pursue terribly far. Descartes and Leibniz, by virtue of their own appeals to notions of the quantitative indefinite, were able to carry these matters further in some regards. But by virtue of the close association (with Descartes) or identification (with Leibniz) of the indeterminate– what they called the indefinite– and the infinite, the work of Descartes and Leibniz also registered a step backward so far as the *prominence* and *independence* of a notion of the parafinite was concerned. What we see in Descartes and Leibniz is a relapsing of the parafinite back into the realm of the infinite. This absorption of the parafinite by the infinite was to have lasting consequences for modern treatments of

---

[16]Blaise Pascal, *Oeuvres Complètes*, ed. Jacques Chevalier (Paris: Bibliothèque de la Pléiade, NRF), 589-91, 1105-07.

quantity, and in this specific regard the problem of large numbers did not reemerge again as such until the beginning of the twentieth century.[17] This later episode is the third scene in the history of the parafinite's long shadow, which we will pick up in the third part of this book.

There is one loose, but critical, end that remains to be tied up. Why did Galileo think that a (finite) continuum could not contain an infinitude of finite parts? We (or many of us, anyway) would now, for example, be perfectly happy to characterize the series

$$\frac{1}{2} + \frac{1}{4} + \frac{1}{8} + \cdots = 1$$

as an infinite summation of finite parts; and, indeed, so already was Leibniz. This infinite sum of fractions is naturally represented in terms of divisions along a line of total magnitude one: first cut the line in half, making two lines of length $\frac{1}{2}$, then cut the second half in half, making two lines of length $\frac{1}{4}$, then cut the second of these in half, making two lines of length $\frac{1}{8}$, and so on. In this way, we divide the line of unit length into a sequence of line segments of length $\frac{1}{2}$, $\frac{1}{4}$, $\frac{1}{8}$, and so on. If we take all of these line segments together, we reconstitute the original line of length one. To all appearances, this seems to give us a geometric picture of the sum

$$\frac{1}{2} + \frac{1}{4} + \frac{1}{8} + \cdots = 1$$

and nowadays we (or most of us, anyway) wouldn't have any qualms about seeing this as a sum of infinitely many finite quantities (namely, $\frac{1}{2}$, $\frac{1}{4}$, $\frac{1}{8}$, etc.) to a finite quantity, namely 1.

So what is the problem here? From the perspective of the Galilean parafinite we can say that characterizing this sum as an

---

[17]For a first indication of how these issues emerge in the context of early twentieth century debates in the foundations of mathematics and in metaphysics, see J. Alberto Coffa, *The Semantic Tradition from Kant to Carnap: To the Vienna Station* (Cambridge: Cambridge University Press), esp. 102.

infinite summation of finite parts begs the question, since in the example given above the sizes of the quantities included must be taken to include quantities as small as you please, and the number of quantities summed must be understood to be more than any given finite quantity, i.e. as many as you please. So it is begging the question to say that this is an infinite summation of finite quantities. From Galileo's perspective it is very clearly a sum of (collectively) parafinite quantities, i.e. from this collection we may choose quantities as small as we wish, and if we were to divide a continuum of unit length into (disjoint) parts of length $\frac{1}{2}$, $\frac{1}{4}$, $\frac{1}{8}$, etc., the number of parts would clearly be, from Galileo's perspective, parafinite, not finite or infinite. Although he does not *explicitly* affirm it, neither does Galileo deny that a parafinite number of finite quantities can sum to a finite quantity, and the affirmation is in any case consonant with his position. Further, although any two quantities chosen from the collection would stand in some finite ratio, we could always find ratios as large (or as small) as we please, and so in this sense we could speak of the collection as a whole as incomparable, since the ratios of terms are as large (or small) as you please, i.e. answering to every designated number, i.e. parafinite.

This is not the end, however, but in fact only the very beginning, of the story. For Galileo's conception of the parafinite as non-infinite rests on his conviction that the parafinite stands at the furthest remove from the infinite. And this conviction, in turn, rests on an argument which assumes that there is an infinite number and then shows that larger and larger numbers become further removed from it. As already mentioned above, Galileo argues that in the case of an infinite number it would have to coincide with all its roots and powers, and since the only number Galileo can find having this feature is unity, he identifies infinity and unity: "...if any number may be called infinite, it is unity. And truly, in unity are those conditions and necessary requisites of the infinite number" (GO, 83).[18]

However, as Leibniz was perhaps the first to recognize explicitly, by just the argument Galileo gives it must also be concluded that infinity contains all its multiples. Just as Galileo argued that in the infinite there are just as many numbers as there are square numbers, or cube numbers, etc., so we can also argue that there are as many numbers as there are doubled numbers, or tripled numbers, or quadrupled numbers, etc. That is, each of the following successions can be aligned with the next:

1, 2, 3, 4, ... (whole numbers)

1, 4, 9, 16, ... (square numbers)

1, 8, 27, 64, ... (cube numbers)

2, 4, 6, 8, ... (doubled numbers)

3, 6, 9, 12, ... (tripled numbers)

4, 8, 12, 16, ... (quadrupled numbers)

But notice that unlike the first three– the whole, square, and cube numbers– the next lines do *not* begin with the number one. And that shows up a limitation in Galileo's argument. Galileo took infinity to be equal to the number one because one contains all its own powers: one squared, i.e. one times one, is one; one cubed, i.e. one times one times one, is one, etc. But one *does not* contain all its own multiples: one times two is two, not one, and one times

---

[18]In this sense, but *only* in this special sense, Galileo is prepared to recognize an infinite number, but it is likely that Galileo would not have recognized unity as infinite *qua* quantity but rather *qua* generative source of number (i.e. quantity). As Josef Hofmann reminds us, "... we ought to remember that 'number' at the time was understood properly to signify only a natural number and that 'one', according to the ancient Pythagorean conception, was not seen as a number at all but rather as the source and origin of all numbers" (Josef E. Hofmann, *Leibniz in Paris 1672-1676: His Growth to Mathematical Maturity* (Cambridge: Cambridge University Press, 1974), 21).

three, is three, not one, etc. Leibniz recognized that this was a serious gap in Galileo's argument, and argued that since there is no number which contains all its powers *and* all its multiples, that the infinite is *nothing*, i.e. zero. For Leibniz, zero was not a number on a par with one, two, three, etc. But since we now think of zero as a number, from our contemporary vantage, we could say that zero *is* a number which contains all its powers and multiples: zero times zero is zero, zero times zero times zero is zero, etc; and zero times two is zero, zero times three is zero, etc. Since for Leibniz zero is not a number but rather *nothing at all*, Leibniz took this as a proof that there is no infinite number![19]

This undercuts Galileo's claim that large numbers are more and more removed from infinite number since there now is no infinite number from which to be removed. More importantly, however, it leaves the position vacated by infinite number open to be filled by just that conception of indefinite quantity which Galileo understood as parafinite. That is, it leaves open a location for the *indefinite as infinite*, rather than as parafinite, and it was in just these terms that Leibniz would understand the infinite in the domain of the quantitative. Galileo wedged the parafinite in between the finite and the infinite. Eliminating the infinite as a quantity, the parafinite could move into this vacant spot and masquerade as a new version of the infinite. With Leibniz's decision to cast the indefinite as infinite rather than parafinite, the parafinite was lost, and the infinite regained.[20]

---

[19]G. W. Leibniz, *Sämtliche Schriften und Briefe* (Darmstadt/Leipzig/Berlin, 1923-), III, 1, 11. This is drawn from the manuscript *Accessio ad Arithmeticam Infinitorum* from 1673, which Leibniz originally intended to submit to the *Journal des Sçavans*. I discuss this manuscript at length in my paper, "Towards Paris: The Growth of Leibniz's Paris Mathematics out of the Pre-Paris Metaphysics," *Studia Leibnitiana*, Band XXXI/2 (1999), 159-180. See also the next chapter.

[20]Thanks especially to Angus Fletcher and Ruby Shattuck for editorial help with this chapter.

Chapter 4

Leibniz on the Indefinite Infinite

4.1 Introduction

Consider the natural numbers: 1, 2, 3, ... Typically, we take there to be infinitely many of them, and, if pressed, we might offer something like the following reasoning. If there were only finitely many natural numbers, then there would be a largest natural number. But if we denote this supposed largest natural number by 'n', then n+1 is also a natural number, contradicting the fact that n is assumed to be the largest natural number. Consequently there are not finitely many, but rather infinitely many, natural numbers.

I don't want to criticize this argument, which in any case is only meant to describe what I take to express some of our informal preconceptions about the nature of the finite and the infinite. Instead, I want to juxtapose to it an argument that Leibniz uses to show that the number of finite (whole) numbers cannot be infinite. The text from which this argument is drawn, "On the Secrets of the Sublime, or on the Supreme Being," dates from early 1676, a period of tremendous intellectual upheaval in the life of Leibniz.

Leibniz was living in Paris and had just invented the infinitesimal calculus.[1] He was engaged in an intensive reading of the Cartesian philosophers, especially Descartes and Malebranche, and was to meet Spinoza later that year on his way back to Hannover, where, apart from extended periods of travel, he would reside for the rest of his life. Here is the argument that Leibniz gives:

> If the numbers are assumed to exceed each other con-
> tinuously by one, the number of such finite numbers
> cannot be infinite, for in that case the number of num-
> bers is equal to the greatest number, which is assumed
> to be finite. It has to be replied that there is no great-
> est number. But even if they were to increase in some
> way other than by ones, yet if they always increase by
> finite differences, it is necessary that the number of all
> numbers always has a finite ratio to the last number;[2]
> further, the last number will always be greater than the
> number of all numbers. From which it follows that the
> number of numbers is not infinite; neither, therefore, is
> the number of units.[3]

---

[1]The term 'infinitesimal calculus' is that most commonly used by Leibniz to refer to his invention, but it is unfortunately misleading, since the fundamental object of the Leibnizian calculus is not the infinitesimal but rather the *differential*. See DLC, esp. 16.

[2]Here Leibniz later added a note: "Rather (N.B.) one proves by this only that a series is endless."

[3]G.W. Leibniz, *De Summa Rerum: Metaphysical Papers, 1675-1676*, trans. G.H.R. Parkinson (New Haven: Yale University Press, 1992), 31-33; further references to be cited as, e.g., (DSR, 31-33). Other works by Leibniz frequently used will be cited as follows: G.W. Leibniz, *New Essays on Human Under-standing*, trans. and ed. Peter Remnant and Jonathan Bennett (Cambridge: Cambridge University Press, 1981); hereafter (NE). *Philosophical Papers and Letters*, trans. and ed. Leroy Loemker (Dordrecht: Reidel, 1969); hereafter (PPL); *Philosophical Essays*, trans. Ariew and Garber (Indianapolis: Hackett, 1989); hereafter (PE); *Mathematische Schriften*, ed. C.I. Gerhardt (Halle, 1849-63, repr. Olms Verlag); hereafter (MS); *Die Philosophische Schriften von*

In this passage, Leibniz distinguishes between there being no greatest number and the number of finite numbers being infinite. That is, Leibniz distinguishes between the indefinite progression of finite numbers, which he here takes to be the case, and the finite numbers being infinite *in number*, which he here takes to be impossible. This is made particularly clear when, directly following the passage cited above, Leibniz goes on to add to the conclusion he has just reached: "Therefore there is no infinite number, or, such a number is not possible" (DSR, 33).[4]

Leibniz's claim may seem odd for two almost antithetical reasons. First, it's hard to imagine that Leibniz is denying that there are infinitely many numbers. The infinitude of natural numbers is something we usually accept nowadays without much question, whether we are simply relying on the fact that there are plenty of them, that they never come to an end, or even that the collection of all of them has infinitely many things in it. But Leibniz is arguing at a fineness of grain for which these three descriptions amount to very different things.

This leads to the second, very different sort of oddness we may find in Leibniz's argument: his distinction cuts so fine that we may worry it makes no difference at all. What, indeed, is the difference between saying that the natural numbers progress indefinitely and saying that there are infinitely many of them? If the thesis of this book makes any sense at all, the answer must be that it makes *plenty* of difference, perhaps more than we could ever imagine. For the first notion of the natural numbers as indefinitely progressing merely commits us to their being *parafinite*, while the second notion commits us to their being genuinely *infinite*.

Eventually, Leibniz will identify the parafinite and the infinite,

---

Gottfried Wilhelm Leibniz, ed. C.I. Gerhardt (Berlin 1875-90, repr. Olms Verlag); hereafter (PS).

[4]The end of this reading ('possible') is conjectural, but in any case the point is clear from the first clause of the sentence.

appropriately construed. What makes this text important is that it shows that in 1676 he hadn't done so yet.

Leibniz explicitly discusses this distinction between the indefinitely progressing and the infinite earlier in the same writing, before reaching the conclusion that an infinite number is impossible. Here, focusing on the infinitely small, Leibniz says:

> One must see if it can be proved that there exists something infinitely small, but not indivisible. If this exists, wonderful consequences about the infinite would follow. Namely: if one imagines creatures of another world, which is infinitely small, we would be infinite in comparison with them.... From which it is evident that the infinite is– as indeed we commonly suppose– something other than the unlimited. (DSR, 31)

Leibniz goes on to say that since the hypothesis of infinites and of infinitely small things is admirably consistent and is successful in geometry, this also increases the probability that they really exist (DSR, 31). As this last remark indicates, the metaphysical outpourings of 1676 should in large part be understood as motivated by the success of the recently invented infinitesimal calculus, which has washed over Leibniz's metaphysics with the force of a tidal wave. Yet as these two passages from "On the Secrets of the Sublime" indicate, the conclusions to be drawn are far from clear. Leibniz has not yet developed a consistent metaphysical position; rather we see his views in flux. But at least throughout the course of these reflections Leibniz retains the belief that the indefinite is to be distinguished from any candidate for an infinite number. Whether infinitely small quantities exist is another matter.

The view that the indefinite is distinct from the infinite was not unknown in the seventeenth century. As we have already seen, in Galileo's *Dialogues Concerning Two New Sciences*, Salviati, speaking for Galileo, develops a conception of the the parafinite. In terms of this conception of the parafinite, Galileo is able to ensure the indefinite division of the continuum without enlisting the traditional Aristotelian distinction between the actual and the potential, and also without committing to the number of parts being either finite or infinite.

Likewise we find Descartes adopting an idea of indefinite progression which is to be identified neither with the finite nor the infinite. In a letter of 6 June 1647, Descartes responds to Chanut concerning the extent of the world:

> In the first place I recollect that the Cardinal of Cusa and many other Doctors have supposed the world to be infinite without ever being censured by the Church; on the contrary, to represent God's works as very great is thought to be a way of doing him honour. And my opinion is not so difficult to accept as theirs, because I do not say that the world is *infinite*, but only that it is indefinite. There is quite a notable difference between the two: for we cannot say that something is infinite without a reason to prove this such as we can give only in the case of God; but we can say that a thing is indefinite simply if we have no reason which proves that it has bounds... Having then no argument to prove, and not even being able to conceive, that the world has bounds, I call it *indefinite*. But I cannot deny on that account that there may be some reasons which are known to God though incomprehensible to me; that is why I do not say outright that it is *infinite*.[5]

---

[5] *The Philosophical Writings of Descartes*, vol. III: The Correspondence, trans. Cottingham, Stoothoff, Murdoch and Kenny (Cambridge: Cambridge University Press, 1991), 319-20. The original letter may be found in *Oeuvres de Descartes*, ed. Adam and Tannery, new presentation in 12 vols. (Paris: Vrin, 1964-76), V, 51-52. This passage is briefly discussed with reference to the doctrine of Nicholas of Cusa in Alexandre Koyré, *From the Closed World*

In contrast to Galileo, Descartes does not commit to whether the world is finite or infinite. Given man's limited rational capacities, for Descartes the indefinite is reducible neither to the one nor the other.

Leibniz's mature philosophical position will be, like Galileo's, that there are indefinitely many parts in the continuum. What distinguishes Leibniz from Galileo is that, modifying the position he took earlier in "On the Secrets of the Sublime," Leibniz will take this to mean that there are *infinitely many* whole numbers, or (finite) parts in the continuum. That is, Leibniz takes the indefinite *as infinite*. My goal in this chapter is to chart the steps Leibniz traversed, and some of the difficulties he encountered, *en route* to this position. I will begin, however, at the end, with an overview of the position Leibniz ultimately achieved.

4.2   The Indefinite as Infinite in Leibniz's Mature Philosophy

In the brief chapter, "Of Infinity," in the late *New Essays on Human Understanding* (1703-05), Theophilus, speaking for Leibniz, provides a survey of the Leibnizian infinite. It is perfectly correct, Theophilus reports, to say that there is an infinity of things, i.e. that there are always more than one can specify (NE, 157). Here Leibniz makes exactly that identification of the infinite and indefinite which Galileo explicitly denies. What makes this identification possible for Leibniz is revealed when Theophilus says, "But it is easy to demonstrate that there is no infinite number, nor any infinite line or other infinite quantity, if these are taken to be genuine wholes" (NE, 157). Leibniz originally arrived at roughly

---

to the Infinite Universe (Baltimore: Johns Hopkins University Press, 1957), 6. Descartes' conception of the indefinite is discussed at length in the fifth chapter of Koyré's book, "Indefinite Extension or Infinite Space," 110-124.

this position in specific opposition to Galileo, having shown the Galilean notion of the infinite to be contradictory, hence impossible. As you may recall from the last chapter, Leibniz showed that infinity cannot be identified with unity, since infinity must contain not only its powers but also its multiples. Unity doesn't contain its multiples, so Leibniz concludes that infinity can only be identified with nothing, i.e. zero, which for Leibniz means that it is impossible. Once this barrier is removed, Leibniz's identification of the indefinite with the infinite first becomes a possibility.

Accepting an infinity of things, but not infinite wholes, amounts to accepting what Leibniz, following tradition, calls a syncategorematic infinite rather than a categorematic one.[6] The distinction between the syncategorematic and the categorematic is standard in the medieval scholastic tradition. The 14th century philosopher and theologian Ockham, for example, says:

> Categorematic terms have a definite and fixed signification, as for instance the word 'man' (since it signifies all men) and the word 'animal' (since it signifies all animals), and the word 'whiteness' (since it signifies all occurrences of whiteness). Syncategorematic terms, on the other hand, as 'every', 'none', 'some',

---

[6]The attempt to use the distinction between the syncategorematic and the categorematic to resolve sophismata pertaining to the infinite is traditional as well. An excellent discussion of one such account, that of Albert of Saxony, is given in Joel Baird's article, "Albert de Saxe et les sophismes de l'infini," in *Sophisms in Medieval Logic and Grammar*, ed. Stephen Read (Dordrecht: Kluwer, 1993), 288-303. Baird focuses on the sophism, "Infinita sunt finita" (Infinites are finite), treated by Albert of Saxony but going back at least to the *De solutionibus sophismatum*, circa 1200 (see Baird, 288). Henri de Gand proves this sophism as follows: "INFINITA SUNT FINITA. Probatio: duo sunt finita, tria sunt finita, et sic in infinitum; ergo infinita sunt finita." (IN-FINITES ARE FINITES. Proof: two are finite, three are finite, and thus to infinity; therefore infinites are finite (for reference, see Baird, 291n.18). A close analogue of this proof recurs in Leibniz's correspondence with Bernoulli, and will be discussed below.

'whole', 'besides', 'only', 'in so far as', and the like, do not have a fixed and definite meaning, nor do they signify things distinct from the things signified by categorematic terms. Rather, just as, in the system of numbers, zero standing alone does not signify anything, but when added to another number gives it a new signification; so likewise a syncategorematic term does not signify anything, properly speaking, but when added to another term, it makes it signify something or makes it stand for some thing or things in a definite manner, or has some other function with regard to a categorematic term.[7]

Syncategorematic terms, then, are terms which have no meaning when they stand alone, but only when they occur inside a larger term or phrase. Leibniz's indefinite infinite is syncategorematic in the sense that the term 'infinite' only signifies when applied to finite numbers, e.g., more than any given finite number. The term 'infinite', standing alone, has no proper meaning, but 'infinitely many' simply means 'more than any given finite number'. In a way, then, it doesn't make any sense to ask what the infinite *is*, since by itself it has no meaning whatsoever.

Another way the point can be put is to say that the sense of the infinite is negative rather than positive. The in-finite is un-limited, and consequently not finite. Leibniz does actually admit a positive sense of the infinite, but not in the domain of number. An indefinite infinite is distinct from what Leibniz calls the true infinite, which, strictly speaking, is only in the *absolute*, which precedes all composition and is not formed by the addition of parts (NE, 157). The *absolute* infinite is a substance, not a number: namely, God. In Leibniz's scheme, God as infinite is neither cat-

---

[7]William of Ockham, *Philosophical Writings: A Selection*, trans. Philotheus Boehner, O.F.M., revised by Stephen Brown (Indianapolis: Hackett, 1990), 51.

egorematic nor syncategorematic, but rather *hyper*categorematic, since God is beyond all terms (PPL, 31 = PS, II, 314).[8] In this way, Leibniz distinguishes the absolute both from the syncategorematic infinite– what we have in the domain of number– and the categorematic infinite– which is impossible. In contrast to the hypercategorematic infinity, the syncategorematic infinity of things is not the true infinite, and our attempts to speak of an infinite number, or an infinite line or other infinite quantity, taken as a whole, result from a confusion of the syncategorematic and the hypercategorematic. To speak of an infinite number would be to confuse the numerical and the substantial and elevate the infinite number to the role of God. In speaking of infinite numerical wholes, we would mistakenly attempt to attribute a true infinity to something which is not absolute. It is *not* absolute, hence not truly infinite, because it has parts, which are limitations. God, on the other hand, has no parts: the absolute is perfect in its lack of limitation. To take an infinite number as a whole, i.e. categorematically, is to attempt to treat a syncategorematic infinity in a way which is only appropriate for such an absolute, hypercategorematic infinite.

By Leibniz's lights, the problem here is that when we try to take the infinite plurality of things as a whole we reverse the true order of precedence: the syncategorematic notion of infinity in fact *derives* from the hypercategorematic sense. This is made clear when Leibniz derives the thought of the syncategorematic infinite from the thought of likeness. Beginning with a straight line, then doubling it, "it is clear that the second line, being perfectly similar to the first, can be doubled in its turn to yield a third line which is also similar to the preceding ones..." (NE, 158). Because the similitude is perfect, that is, unlimited, it is impossible that the process can ever be hindered: the same principle is always applicable (NE, 158). Leibniz sees our thought of the infinite as deriving from the

---

[8]See also the discussion in Loemker's introduction, PPL, 31 and also PPL, 514n.2 and PPL, 541n.21.

thought of likeness. He then indicates that our error in considering an infinite quantity as a whole (categorematically) would be to take something absolute and attribute it to something limited: "The idea of the absolute, with reference to space, is just the idea of the immensity of God *and thus* [my emphasis] of other things. But it would be a mistake to try to suppose an absolute space which is an infinite whole made up of parts" (NE, 158). This last sentence is no doubt directed at Newton, who was committed to the notion of space as the *sensorium Dei*, the sensory manifold of God, which Newton understood to be an infinitely extensible whole made up of spatial parts.

There are many interesting issues surrounding Leibniz's treatment of the hypercategorematic infinite, among them his debates with Samuel Clarke about Newton's views on space. But my concern is not with Leibniz's notion of the absolute as infinite and the implications of this notion. Rather, it is with the question what room is *left* in Leibniz's metaphysics for the distinct idea of an indefinite, or syncategorematic, infinite. If what is absolute is understood as perfect, hence without the limitation of parts, it is clear that a whole made of parts cannot be absolute. But why can't there be an infinite whole made up of parts? It may be that the passages from the *New Essays* cited above amount to a kind of proof that this is impossible. But it's not entirely clear that they are strong enough to count as such a proof, nor that Leibniz took them to have that status. Unless we already have either an argument to show that an infinite whole must be absolute or else some direct proof that an infinite whole made up of parts cannot exist, an infinite whole made up of parts still seems like an option. The point here is that Leibniz's commitment to a hypercategorematic infinite does not by itself explain the exclusion of a categorematic infinite, nor does his exclusion of a categorematic infinite automatically make room for a syncategorematic one. We need to return to Leibniz's proof of the impossibility of a categorematic infinite–

the one we already saw in the previous chapter– and think about it in even more detail.

4.3   The Status of Infinitesimals

Leibniz originally arrived at the position that there is no largest number, or number of all numbers, in specific opposition to the position of Galileo, during an intensive reading of the *Two New Sciences* in 1672 or 1673. We know that Leibniz was aware of the Galilean notion of the parafinite from notes he made during this reading. In particular, Leibniz notes that to the question whether the parts of the continuum are finite or infinite, Galileo responds that they are neither, but are rather more than any given number. Here Leibniz appends in parentheses: "or indefinite."[9] Hence while Leibniz rejects as impossible what Galileo takes to be the infinite, he will nonetheless appropriate the Galilean parafinite, what he will call the indefinite, and ultimately understand this *as* infinite.

Leibniz's position that there is no number of all numbers is closely related to his declaration in the *New Essays* that there are no infinite wholes. As we have seen in the previous chapter, in the dialogue of the first day of the *Two New Sciences*, Galileo recognizes that the infinite number of all (whole) numbers is equal in size to a proper subcollection of itself, namely the subcollection of all square numbers. Since the same holds for all other powers as well (cube numbers, etc.), Galileo concludes that the infinite number must therefore be whatever number contains all its powers within itself. The only such number is one, or unity, from which indeed all other number is generated, and so Galileo takes the infinite number of all numbers to be one (GO, 78-85).[10]

---

[9]Leibniz, *Sämtliche Schriften und Briefe* (Darmstadt/Leipzig/Berlin, 1923–), VI, iii, 168.

[10]It is important to note that Galileo has Salviati repeatedly deliver a series of disclaimers along with his remarks about the infinite. At (GO, 73) Salviati

Leibniz first responds in detail to Galileo's identification of infinity with the number one in the "Accessio ad Arithmeticam Infinitorum" of 1673. As we've already seen, Leibniz drives Galileo's point even further and notices a host of properties of the infinite whole which prevent it from being identified with unity. The only number which satisfies all these requisites is not the number one but rather zero: zero squared is zero, zero cubed is zero, etc., but also two times zero is zero, three times zero is zero, etc. Given Leibniz's understanding of the number zero, this means that the infinite is nothing, i.e. no number at all: "thus the infinite is impossible, not one, not the whole, but nothing. Thus the infinite number = 0."[11] Since the infinite number is nothing, this means an infinite number is impossible.

Leibniz interpreted his impossibility proof as a consequence of the so-called Euclid axiom that the whole is greater than the part. In other words, if you can take something and put it in correspondence with a proper part of itself, then you weren't talking about a *thing* in the first place. In particular, the natural numbers, 1, 2, 3, ... cannot be taken as a whole: they do not constitute a categorematic infinity. If something is indeed a whole, it must always be greater than any of its (proper) parts.

Leibniz regarded this axiom as itself demonstrable in a syllogism taking as its first premise a definition (of greater) and as its second premise an identical proposition (a part is equal to a part of the whole). Calling the whole *cde* and the relevant part *de*, the

---

says he is "going to produce a fantastic idea of mine which, if it concludes nothing necessarily, will at least by its novelty occasion some wonder." In another passage at (GO, 83) Salviati speaks of "marvels that surpass the bounds of our imagination, and that must warn us how gravely one errs in trying to reason about infinites by using the same attributes that we apply to finites; for the natures of these have no necessary relation between them." See also passages at (GO, 96, 105).

[11]Leibniz, *Sämtliche Schriften und Briefe* (Darmstadt/Leipzig/Berlin, 1923–), II, i, 226.

syllogism runs:

> That whose part is equal to the whole of another is greater, by definition.
>
> A part of the whole *cde*, namely *de*, is equal to the whole *de*, namely itself.
>
> Thus *cde* is greater than *de*, and so the whole [is greater than] the part.[12]

Distilling this, the argument runs:

I That is greater whose part is equal to the whole of another.

II *de* is a part of the whole *cde*.

III Therefore, *cde* is greater than *de*.

To put this in the form of a simple syllogism, I've suppressed the premise which explicitly recognizes that *de* is the whole of *de*. So in fact the argument is a bit more complicated than a simple syllogism, but the main point is retained in the syllogistic form of the argument.

Leibniz was already committed to this argument for Euclid's axiom before his reading of Galileo in 1672 or 1673,[13] and we can best see Leibniz's disagreement with Galileo as a function of his commitment to Euclid's axiom. In the *Two New Sciences*, Galileo had asserted that "the attributes of greater, lesser, and equal do not suit infinities, of which it cannot be said that one is greater, or less than, or equal to, another" (GO,78). While Leibniz draws the conclusion that an infinite number of all numbers is absurd since its existence would contradict a universal axiom, Galileo concludes instead that such an axiom does not pertain in all cases.

---

[12]idem, VI, ii, 482-83.

[13]The syllogistic derivation of this axiom given here dates from 1671, prior to Leibniz's departure for Paris.

In other contexts, Leibniz gives other arguments to show the absurdity of infinite wholes, and we have seen one such at the beginning of this chapter.[14] But Leibniz's syllogistic demonstration seems to continue to be the predominant argument behind his assertion that an infinite whole such as the number of all numbers is impossible. In particular, it is this argument which Leibniz sketches in the context of one of his most intensive debates regarding the nature of the infinite in correspondence with his mathematical protégé, Johann Bernoulli. The heart of Leibniz's debate with Bernoulli lies in the problem of the status of infinitesimals, i.e. infinitely small magnitudes.[15] Do such infinitely small magnitudes exist? Is the idea of such infinitely small magnitudes even conceptually coherent? These are the questions on the table in the debate, but the debate itself is framed by the impossibility of a maximal or minimal magnitude, i.e. a largest or smallest whole. In a letter from August/September 1698 to Bernoulli, Leibniz affirms his commitment to the universal scope of Euclid's axiom. Consequently,

it appears to me that we must say either that the infi-

---

[14]In those contexts in which Leibniz wishes to suppress the more subtle, and technical, issues surrounding the consideration of the infinite, he often substitutes for the proof that there is no greatest number a proof that a fastest motion is absurd. For example, in the *Meditations on Knowledge, Truth, and Ideas*, Leibniz proves the absurdity of a fastest motion as follows: "For let us suppose some wheel turning with the fastest motion. Everyone can see that any spoke of the wheel extended beyond the edge would move faster than a nail on the rim of the wheel. Therefore the nail's motion is not the fastest, contrary to the hypothesis" (PE, 25) = (PS, IV, 424). This proof is closely related to debates about what would happen were space to be limited. For a discussion in a variety of early modern contexts, see Koyré, *The Open Universe*, op. cit., passim.

[15]Ultimately, these issues bear on the existence or nonexistence of infinitely large and small *quantities* too, as the following quotation from Leibniz makes clear. But I will focus on magnitudes here since the debate about infinitesimals is first and foremost a debate about geometric magnitudes.

nite is not truly one whole, or else that if the infinite is
a whole, and yet is not greater than its part, then it is
something absurd. Indeed I demonstrated many years
ago that the number of the multitude of all numbers
implies a contradiction if it is taken together whole.
The same [holds] for a maximum number and a minimum number, or fraction smaller than all others. This
must also be said about a fastest motion and all similar
things. (MS, III, 535)

But, according to Leibniz, this does not by itself rule out the possibility of infinitesimals and infinitely large things, since a maximum
is different from the infinite and a minimum from the infinitely
small (MS, III, 536). He reports to Bernoulli that he cannot determine whether such infinite and infinitely small magnitudes are
possible, and so he will allow the matter to remain in the middle.
He does assert, however, that if these magnitudes can be demonstrated to be possible, then their existence will follow (MS, III,
536).[16]

Leibniz and Bernoulli's debate about the status of infinitesimals
is important because it tells us a great deal about how Leibniz was
thinking of the infinite– in this case the infinitely small– in the
realm of magnitude. In the 1670's, after Leibniz had established his

---

[16]The situation regarding the existence of infinitesimals bears considerable
structural analogy to the situation concerning the ontological proof of the
existence of God. Leibniz criticized the Cartesian ontological argument on
the grounds that it merely proves that *if* God's existence is possible, then it
is actual, but it does not give a proof of the *possibility* of God's existence.
Leibniz was much preoccupied with such a proof of the possibility of God's
existence during 1676, and returns to it later in writings of 1678, the mid-
1680's, and 1714. The issues involved in the provision of such a proof are
extremely intricate and consequently well beyond the bounds of this chapter.
The reader is referred to the detailed discussion in Robert Merrihew Adams'
Leibniz: Determinist, Theist, Idealist (Oxford: Oxford University Press, 1994).
See esp. 141ff.

view that infinite wholes are impossible, he still seems inclined, at least sometimes, toward the existence of infinitesimal magnitudes–as, for example, in the passage from 1676 quoted above. But what is most important in the debate with Bernoulli is that *Bernoulli* thinks that Leibniz is *already* committed to infinitesimals by virtue of the way he understands the infinite. It is around this issue that the debate, or at least the portion of it I want to focus on, revolves.

Bernoulli believes that a commitment to the existence of infinitesimals follows from Leibniz's assertion that the physical continuum of bodies is actually divided into infinitely many parts. He offers Leibniz the following argument in support of his contention:

> Consider any determinate magnitude divided into parts according to this geometric progression: $\frac{1}{2} + \frac{1}{4} + \frac{1}{8} + \frac{1}{16} +$ etc. As long as the number of terms is finite, I confess the singular terms will be finite; but if all the terms *actually* exist there will surely be infinitesimals and all of the following infinitely small magnitudes. (MS, III, 529)

Bernoulli goes on to point out that in bodies all such divisions are (according to Leibniz) actual, so that, Bernoulli argues, infinitesimals, i.e. infinitely small magnitudes, would be necessary.

Like the position Leibniz took in 1676 in "On the Secrets of the Sublime," Bernoulli's position is apparently that there is *not* an infinite number of finite terms in the series $\frac{1}{2} + \frac{1}{4} + \frac{1}{8} +$ etc; he reaches this conclusion on the basis of an argument which runs as follows: suppose that a finite portion of matter is already actually divided into an infinite number of parts and yet none is forced to be infinite (by which Bernoulli presumably means no individual part is infinitely small; this is at any rate how Leibniz understands him). Then the single parts are finite, and if the single parts are finite, then all of them together make up an infinite magnitude, contrary to hypothesis (MS, III, 529).

Despite Bernoulli being an accomplished practitioner of the infinitesimal calculus, it may perhaps be tempting to dismiss this as a quaint fallacy generated by an archaic (mis)understanding of the nature of the mathematical infinite. This would be a mistake (and, in fact, there is a modern approach to the mathematics of infinitesimals that supports Bernoulli's concerns extensively).[17] Bernoulli's position was expressive of a pervasive worry in early modern discussions of the infinite, and we've already seen an almost identical argument occurs in Galileo's *Two New Sciences*. Indeed, Galileo not only faced a similar problem but resolved it without recourse to the infinite by understanding the number of terms in such a series to be parafinite, i.e. as many as you please. This makes Leibniz's response to Bernoulli a privileged locus for seeing how Leibniz identifies the indefinite with the infinite rather than, as Galileo had, taking the indefinite to be a *sui generis* quantity all its own.

Leibniz responds to Bernoulli's example by admitting that if there were a finite term no smaller than every term in the infinite series then the sum would indeed be infinite.[18] Suppose, for example, that all terms in the series are at least $\frac{1}{2}$. Then the sum of the series would be greater than $\frac{1}{2} + \frac{1}{2} + \frac{1}{2} + \ldots$, and this is clearly infinite. But Leibniz does not grant Bernoulli's conclusion in general. In particular, when the terms in the series, while all finite, become smaller and smaller, tending toward zero, Leibniz

---

[17]For an exposition of this approach to non-standard arithmetic (and, following on that, non-standard analysis), see Alain M. Robert, *Nonstandard Analysis*, trans. and adapted by the author, reprinted (New York: Dover Publications, 2003), esp. 11. Bernoulli's approach is not vindicated by the better known Robinsonian approach to non-standard analysis; for a comparison, see Robert, 30. I discuss these matters in more detail in the final Envoi.

[18]This is of course a sufficient, but not a necessary condition, as Leibniz was well aware. Already in 1673 Leibniz took the series $\frac{1}{1}+\frac{1}{2}+\frac{1}{3}+\frac{1}{4}+\ldots$ to diverge. For a discussion see Joseph E. Hofmann, *Leibniz in Paris 1672-1676: His Growth to Mathematical Maturity* (Cambridge: Cambridge University Press, 1974), 21.

does not grant that the sum will *always* be infinite– although it will in some cases.[19] Leibniz replies that, "even if we take all the terms in the progression $\frac{1}{2}$, $\frac{1}{4}$, $\frac{1}{8}$, ... etc. actually to exist, I do not hold anything to follow hence except that there are actually given finite assignable fractions of any given smallness" (MS, III, 536). This is precisely an indefinite collection of fractions, like Galileo's parafinite. Leibniz makes recourse to this indefinite but thinks of it as an infinitude of terms, and insists this is not in conflict with the infinitude of parts actually being given.

Unlike Galileo, Bernoulli goes on in a later letter (6 December 1698) to say that between the finite and the infinite no third term is given (MS, III, 555), and in the same letter he asserts that either all the terms in the series are not *actually* given, and then only finitely many are given and more could be given, or else all the terms are *actually* given, and then there are an infinite number of them, hence infinitesimals. In the former case it is clear that the number of finite terms is finite; but in the latter case as well, it seems, the number of *finite* terms must still be finite. For otherwise, according to Bernoulli, the series would sum to an infinite magnitude, which is contrary to hypothesis. What, then, in this latter case, could this number of finite terms be? Is it some determinate finite number, or is it some (indeterminate) finite number larger than any finite number we can supply?

For those of us who are comfortable summing infinite series, like

$$\tfrac{1}{2} + \tfrac{1}{4} + \tfrac{1}{8} + \dots,$$

Bernoulli's turn to the issue of infinite number serves as a useful prod. Consider, for example, Leibniz's position as stated in *The Monadology* that there is an infinitude of monads, or simple substances. According to Leibniz's syncategorematic notion of the indefinite as infinite, this means that the number of monads is greater

---

[19]See the example in the previous note.

than any given (finite) number. But these monads are actual substances! Shouldn't there be a *number* of them?! Bernoulli reasons that if Leibniz is forced to admit an infinite number he should not bristle at having to admit infinitesimal magnitudes. Leibniz may be able to defuse Bernoulli's parry in this specific mathematical circumstance by explaining why he isn't admitting an infinite number, but what about when we're talking about the number of *monads*? In such a case it seems particularly odd to say that there isn't a finite number of them but there isn't an infinite number of them either! Monads are actually existing substances: how could there fail to be a number of them?

In any case, the upshot is that, for Bernoulli, there is no room to admit Leibniz's indefinite infinite, which seems to Bernoulli to have the status of an impossible third thing between the finite and the infinite. Since Leibniz rejects Galileo's conception of the infinite as a largest or absolute magnitude, nothing blocks him, as it does Galileo, from recognizing the indefinite as infinite; Leibniz thus avoids the need to posit a middle term between the finite and the infinite, as Galileo did. Yet neither does the gap left by Leibniz's proof of the impossibility of an absolute magnitude in any positive way *sanction* his declaration that the indefinite *is* infinite. Surprisingly, from our own hypermodern perspective,[20] Bernoulli remains unrefuted!

4.4 The Consolidation of Leibniz's Position

In the correspondence with Bernoulli, there is a potentially competing candidate for the role played by the indefinite infinite in both the realms of the infinitely large and small: although he can-

---

[20]I.e., recognizing the interpretation of Nonstandard Analysis given by Robert and the so-called Reeb school in general.

not demonstrate their existence, Leibniz also cannot find a way to rule out infinitesimal (but not minimal) and infinitely large (but not maximal) magnitudes. If such were to exist, we may ask, what would the status of Leibniz's indefinite infinite be? Would it be supplanted by these true infinitesimal and infinite magnitudes? By 1703/1705, the period during which Leibniz wrote the *New Essays*, Leibniz has moved to the position that *any* infinite or infinitesimal magnitude is impossible. In the *New Essays* he says, for example:

> But it would be a mistake to try to suppose an absolute space which is an infinite whole made up of parts. There is no such thing: it is a notion which implies a contradiction; and these infinite wholes, and their opposites, the infinitesimals, have no place except in geometrical calculations, just like the use of imaginary roots in algebra. (NE, 158)

Here Leibniz takes the opposite of infinite wholes to be infinitesimals and declares neither to exist.[21] What specifically motivates Leibniz to change his position in the period between 1699 and 1705 is, I believe, his continuing engagement in debates concerning the status of infinitesimals. But what most fundamentally drove these changes is Leibniz's deepening sense of his commitment to the indefinite as infinite. In order to point out some of these developments I would like to turn to one of the central documents regarding the status of infinitesimals which Leibniz composed during this intervening period.

---

[21]It might be objected that what Leibniz refers to here is not infinitely small magnitudes, but minimal magnitudes in opposition to maximal wholes. This interpretation is, I believe, extremely implausible, however, because Leibniz goes on to speak of just these infinite wholes and their infinitesimal counterparts as what *do* have a place in geometrical calculations *sub specie imaginationis*. I believe the conclusion is inescapable that Leibniz is here declaring the impossibility of the existence of infinitesimals.

In a letter to Varignon from 2 February 1702, along with a short note on the "Justification of the Infinitesimal Calculus by That of Ordinary Algebra" (PPL, 542-546) = (MS, IV, 91-5, 104-6), Leibniz's position regarding infinitesimals is the negative one that we need not be committed to infinitesimals in any rigorous metaphysical sense. In this letter, Leibniz pleads that it is unnecessary to make mathematical analysis depend on metaphysical controversies or to make sure that there are lines in nature which are infinitely small in a rigorous sense in contrast to our ordinary ones... (PPL, 542-3) = (MS, IV, 91). Leibniz's focus is on the lack of a logical implication *from* the lack of a finite whole *to* an infinitely large or small quantity– even as he fails to rule such an implication out. It is precisely the *unnecessariness* of such an implication which permits the free conduct of mathematics without an engagement in metaphysical disputes about the existence of the infinite. Not only do we not need to have a proof that infinite wholes are contradictory in order to make mathematical use of the infinitesimal calculus, equally we do not need to decide whether infinitely large and small quantities exist or not.[22] Here Leibniz is most interested in gaining support for the weaker position that such infinite wholes are *unnecessary* (as are infinitely large and small magnitudes) for the proper conduct of mathematics. On this weaker position, neither is mathematically or metaphysically requisite for dealing with infinite pluralities.[23]

---

[22] This strategy of arguing for a relative independence of a given area from metaphysical intrusion while maintaining nonetheless the fundamental need for a proper metaphysical foundation is a pervasive one in Leibniz's philosophy, with the emphasis on independence generally increasing during his later years. On this issue, see my 1995 University of Chicago dissertation, *Labyrinthus de Compositione Continui: The Origins of Leibniz' Solution to the Continuum Problem*, 149-151.

[23] Even in pieces post-dating the *New Essays,* Leibniz often emphasizes this weaker position; but even in such contexts he demonstrates more definite commitments than he does in the passages considered above from the correspon-

But in a letter which Leibniz sent to Varignon in 1702, he indicates that he thinks he in fact has a *proof* that infinitesimals are impossible (MS, IV, 110). As early as the Paris period, when Leibniz first invented the calculus, he wrote a long essay in which he showed that, at least in the context of certain central mathematical problems, the application of the calculus could be rigorously justified.[24] What Leibniz is really after in the letter to Varignon, on the other hand, is a way to *understand* the infinitesimal calculus, not just to justify it. It is in the letter to Varignon and writings from the same time that Leibniz spins out such an understanding in terms of the idea of finite quantities standing in a relation of incomparability. To make the notion of such incomparable quantities vivid, he draws on physical ideas of incomparability to illustrate this notion:

> It is in this sense that a bit of magnetic matter which passes through glass is not comparable to a grain of sand, or this grain of sand to the terrestrial globe, or the globe to the firmament. (PPL, 543) = (MS, IV, 91-92)

This provides a picture in terms of which we may think of working with the infinite without appealing to infinite magnitudes. In doing so, Leibniz makes it clear just how weak a commitment to the infinite is required to work in the mathematical realm, even in the infinitesimal calculus: we might say that the calculus is rather naturally a parafinite theory. Instead of relying on quantities as large or small as we wish, that is, quantities bearing as small or

dence with Bernoulli. See, e.g., (PE, 267) = (PS, VI, 592).

[24]See G.W. Leibniz, *De quadratura arithmetica circuli ellipseos et hyperbolae cujus corollarium est trigonometria sine tabulis*, ed. Eberhard Knobloch (Göttingen, Vandenhoeck & Ruprecht, 1993); an edition with French translation with commentary is provided in G.W. Leibniz, *quadrature arithmétique du cercle, de l'ellipse et de l'hyperbole*, ed. Eberhard Knobloch, trans. with introduction and notes by Marc Parmentier (Paris: Vrin, 2004).

large as we wish a (finite) ratio respectively to some fixed quantity, Leibniz instead opts to consider quantities which are taken to have *no* finite relation: we take them, instead, to be incomparably greater or smaller than ours (PPL, 543) = (MS, IV, 91). Leibniz is effectively saying that they are *not* in any finite relation to each other, but he stops short of saying that they *are* in an infinite ratio to each other in any positive sense. Leibniz is working with symbolic magnitudes (although it is necessary to ask exactly what it means to speak of them *as* magnitudes) which are stipulated *not* to have relations of proportion with the magnitudes that we usually employ. These magnitudes, then, are representatives of the fact that the failure of a finite ratio to obtain need not imply that an infinite ratio positively does obtain; logically they stand, so to speak, half way between the one and the other. Further, they are representatives of the fact that no such infinite ratio is *needed* to conduct the infinitesimal calculus: *there are, or at least need be, no infinite or infinitesimal ratios in the infinitesimal calculus.* These strange finite, yet incomparable, magnitudes anticipate the notion of scaling which I will introduce toward the end of this book as a fundamental tool for dealing with the mathematics of the parafinite.

Leibniz goes on to show in a note on the "Justification of the Infinitesimal Calculus by Ordinary Algebra" which he sent to Varignon at the same time that there are always finite magnitudes which stand in the same proportions to each other as do those incomparable magnitudes for which they are, so to speak, assigned to be proxies. Thus Leibniz's recourse to these incomparable magnitudes is exclusively symbolic: failing a proof that infinite magnitudes exist, they are mathematical symbols whose true interpretation can only be given in terms of their replacement, in any given context, by other, relatively more free-standing symbols. This makes Leibniz's syncategorematic presentation of the mathematical infinite the natural one. The infinitesimal calculus, as advertised in the title to the note, is justified by ordinary algebra. Beyond the use of certain auxiliary symbols, there appears to be no commitment to the infinite in the infinitesimal calculus. This makes the title of Leibniz's proposed, but never drafted, summa of the infinitesimal calculus, *De Scientia Infinitii*, "On the Science of the Infinite," ring with a certain irony.[25] The infinitesimal calculus is as much, or as little, about infinitesimals as the business of a variety shop is to sell varieties. Consequently Leibniz will later declare that infinitesimals and infinitely large magnitudes are convenient fictions.[26]

What we see here is an instance of Leibniz's commitment to the infinite in the realm of mathematical and physical quantity in the weakest sufficient sense required, and it is in this light that Leibniz's commitment to the indefinite infinite should be understood. This commitment is in fact so weak that we may wish to understand it as no commitment to the *infinite* at all. To be sure, Leibniz is committed to infinite pluralities, such as the terms of an infinite series.[27] But such infinities are indefinite or, to fall back on the Scholastic terminology to which Leibniz has recourse, syncate-

---

[25]Pierre Costabel has tried to explain why this project was never drafted in his "De Scientia Infinitii," in *Leibniz, aspects de l'homme et de l'oeuvre* (Paris, 1966), 105-117. Although I would tend to agree with Costabel that the greatest barriers to this project were of a practical (and perhaps, as Costabel also suggests, psychological) nature, I would also suggest that there may also have been metaphysical reasons why this project was never pursued past the preliminary stages, much less completed.

[26]For example, in a letter from 1716 to Samuel Masson: "The infinitesimal calculus is useful with respect to the application of mathematics to physics; however, that is not how I claim to account for the nature of things. For I consider infinitesimal quantities to be useful fictions" (P, 230) = (PS, VI, 629).

[27]In the letter to Varignon, Leibniz uses this to defend his calculus as a calculus of the infinite: "Yet we must not imagine that this explanation debases the science of the infinite and reduces it to fictions, for there always remains a 'syncategorematic' infinite, as the Scholastics say" (PPL, 543) = (MS, IV, 93).

gorematic. In hypermodern parlance, they are parafinite. So if we ask how many terms there are in an infinite series, the answer is not: an infinite number (if we take this either to mean a quantity which is infinitely larger than a finite quantity or a largest quantity) but rather: more than any given finite quantity. The force of this maneuver is to restrict the infinite, in the realm of the quantitative and consequently limited, to an adjectival role, and this will be precisely Leibniz's point in the chapter on infinity in the *New Essays*: "The idea of the absolute, with reference to space, is just the idea of the immensity of God and thus of other things" (NE, 158). The immensity of God can be *referred* to space; we thereby acquire the idea of space *as* infinite, in the sense of being unending. Infinite space is unlimited space; on the other hand, to elevate the unlimited directly from an adjectival to a substantival level would be to have an infinitely large magnitude, if indeed this were to be possible.

There is yet another indication in the letter to Varignon that Leibniz is moving in the direction of denying infinitesimals outright as well. Toward the end of the letter Leibniz describes the relation between the operation of the finite and the infinite in a way which bears an interesting resemblance to Leibniz's many descriptions of pre-established harmony which he expresses in his philosophical writings:

> Yet one can say in general that though continuity is something ideal and there is never anything in nature with perfectly uniform parts, the real, in turn, never ceases to be governed perfectly by the ideal and the abstract and that the rules of the finite are found to succeed in the infinite– as if there were atoms, that is, elements of an assignable size in nature, although there are none because matter is actually divisible without limit. And conversely the rules of the infinite apply to the finite, as if there were infinitely small metaphysical

beings, although we have no need of them, and the division of matter never does proceed to infinitely small particles (PPL, 544) = (MS, IV, 93-4).

In this passage, Leibniz describes two counterfactuals: we have rules which allow us to treat matter *as if* it were extended yet indivisible and rules which allow us to treat finite extension *as if* it were composed of infinitely small extended metaphysical beings. But in fact the only matter there is is *divisibly extended* and the only metaphysical beings there are are *indivisibly unextended*. This picture effectively leaves no ground for a third *thing* which could mediate metaphysically between these two: like Descartes, Leibniz takes the distinction between the extended and the unextended to be categorical and complete. We are left with nothing more than the phraseology of 'as if', which is our way of understanding the pre-established harmony of the ideal and the real. My point here is not so much logical as it is architectonic: existing infinitesimals, whether ideal or real, fit nowhere in Leibniz's scheme. There is no third, independent realm to which this level of structure could be assigned. The ultimate consequence of Leibniz's understanding that in the realm of quantity there are no infinite or infinitesimal wholes is that the quantitative infinite is *identified* with the indefinite or unlimited. Consequently, the proof which was first taken to prove the impossibility of maximal or minimal units may now be taken more broadly to show the impossibility of *all* infinite or infinitesimal wholes. In a letter to Varignon shortly following the one just discussed, Leibniz in fact reports that he believes he can prove that there are not, nor could there be, any infinitely small things (MS, IV, 110). In Appendix C, I consider such an impossibility proof which Leibniz probably drafted much earlier in his career; it is the only extant attempt at such a proof I know of which has been found in Leibniz's papers.

As Leibniz declared in a letter to the Princess Electress Sophie in 1696, "My fundamental meditations circle around two things,

116

namely unity and infinity" (PS, VI, 542). Leibniz ultimately recognized that in the realm of quantity infinity could in no way be construed as a unified whole. As we have seen, the cleft which this impossibility established between these two most fundamental of Leibnizian concerns left at least one of them with a particularly delicate, if not indeed problematic, status. In Leibniz's philosophy, the parafinite drapes itself in the garb of the indefinite as infinite.[28]

---

[28]This chapter is based on my earlier article, "Leibniz on the Indefinite as Infinite," *The Review of Metaphysics* **51** (1998): 849-874.

Scene Three

More Recent History: Husserl, Russell, Wittgenstein

Chapter 5

Hrusserll and the Status of the Parafinite

It seems impossible to remember everything at once.

—Bertrand Russell[1]

5.1 Introduction: Coffa's Passing Remark

In *The Semantic Tradition from Kant to Carnap: To the Vienna Station*, J. Alberto Coffa remarks, seemingly in passing, that

> The analogy between Russell and Husserl's problems circa 1900 is quite striking. To begin with, mathematics provided in both cases the original stock of problems that led them to what both called logical investigations. Next, within mathematics itself, the foundational problems that initially seemed to trouble them

[1]Bertrand Russell, *Foundations of Logic 1903-1905: The Collected Papers of Bertrand Russell, vol. 4* (London: Routledge, 1994), 257 (from the 1904 set of working notes, Fundamental Notions).

most involved the representation of numbers and, in particular, the difficulty of representing the very large ones.[2]

Coffa's focus on the status of large numbers finds a strong echo in a remark made by Jan Mycielski in his paper, "The Meaning of Pure Mathematics," when he asserts that

> the detachment of certain mathematical objects (such as 10^{100}) from physical objects was the main problem which stimulated Hilbert's and Poincaré's writings in the philosophy of mathematics.[3]

Mycielski's remark provides supporting evidence for the broader claim that the problem of the status of the parafinite occupied at this time a pivotal position with regard to those issues at the intersection of philosophical and mathematical concern.[4] After Galileo's preliminary discovery of the parafinite, and Leibniz's absorption of it into the indefinite as infinite, the status of large numbers among Husserl, Russell, Hilbert and Poincaré marks another locus of especial importance, in which the parafinite once again breaks out from behind its long shadow. My goal in this chapter will be to suggest some reasons why this problem of the status and representation of the parafinite occupied a focal position in the writings of Husserl and Russell, or 'Hrusserll' for short.[5] I will begin this task by piecing together what can be gathered from a reading of the

[2] J. Alberto Coffa, *The Semantic Tradition from Kant to Carnap: To the Vienna Station* (Cambridge: Cambridge University Press, 1991), 102; hereafter cited internally as STKC.

[3] Jan Mycielski, "The Meaning of Pure Mathematics," *Journal of Philosophical Logic* **18** (1989) 315-20; here 316.

[4] Here see also Kant's remark at B16 in the *Critique of Pure Reason*, that the synthetic nature of additive propositions is made more aware to us if we consider the case of large numbers.

[5] I am indebted to Sam Ubrec for this amalgamation, with the express understanding that he should not be held responsible for my use of it in this

larger context within which Coffa situates his remark, and then I will provide a conceptual apparatus for handling the discussion of the parafinite; only after this will I turn to the work of Hrusserll circa 1900.

Most specifically, Coffa's remark occurs in the context of a discussion of Husserl's treatment of the representation of number in his *Philosopy of Arithmetic* of 1891: Husserl's solution to the problem of arithmetical knowledge was to say that it often relies on symbolic representation (STKC, 102).[6] In particular, this is the case whenever we deal with knowledge concerning *large* numbers. Coffa goes on to note that a considerable amount of the *Philosophy of Arithmetic* is devoted to an explanation of what these representations are and how they work, but that this explanation collapses as soon as one distinguishes between psychology and logic, as Husserl would endeavor to do a few years later (STKC, 102). Thus Husserl's attempted solution to the problem of representing the parafinite fell prey, according to Coffa, and indeed according to Husserl's own later self-diagnosis, to psychologism, the conflation of psychology and logic.

For Russell, on the other hand, Coffa takes the main problem to be with the status of the infinite rather than the finite-but-large: one of the central problems in Russell's early philosophy of mathematics was to account for our ability to have knowledge concerning infinite numbers, especially that knowledge recently uncovered by Cantor in his theory of the transfinite (STKC, 102). Coffa implies that Russell found himself in a more fortunate situation than

---

context; for a precedent, see Lewis Carroll's preface to *The Hunting of the Snark* in *The Humorous Verse of Lewis Carroll* (New York: Dover Publications, 1960), 267-9 (thanks to Joshua Striker for this reference).

[6]Husserl declares that the bulk of the work for the *Philosophy of Arithmetic* was already done during 1886-87; see Eley's introduction to *Philosophie der Arithmetik*, pp. xxvi-xxvii, where he quotes from Edmund Husserl, "Persönliche Aufzeichnungen," ed. W. Biemel, in *Philosophy and Phenomenological Research* **XVI** 1956, 294.

Husserl by virtue of an equally marked difference between their positions, namely, that Russell's strategy for dealing with his foundational problems was, from the beginning, to focus on the target of all epistemic attitudes, the proposition (STKC, 102-3). Coffa thinks this was a significantly different, and ultimately more successful, route toward addressing the nature of what might be called improper representations,[7] i.e. a form of representation that was "very unlike" what it represented (STKC, 102).[8]

It is striking how consistently Frege, in *The Foundations of Arithmetic*, uses examples of large numbers to deflate psychological issues of representation and turn to logical ones.[9] Frege begins by asserting that "it is awkward to make a fundamental distinction between small and large numbers, especially as it would scarcely be possible to draw any sharp boundary between them," and goes on to use the problem with the "observational status" of sufficiently large numbers to dispute Mill's mathematical empiricism.[10] Nonetheless, Frege insists that statements concerning large numbers nonetheless have a *sense*, and this fuels the drive away from psychologism and toward logicism. If arithmetic were no more than psychology, "10^{10}, perhaps, might be only an empty symbol, and there might exist no idea at all, in any being whatever, to answer to

---

[7]This term derives from the work of Husserl's teacher, Brentano, whose work lay behind Husserl's approach to representation in the *Philosophy of Arithmetic*; see STKC, 101.

[8]Note that in the discussion of the Prologue, above, the representation of number proceeds from less improper notations (e.g., stroke notation) to progressively more improper ones (Roman numerals, decimal notation, arrow notation). As we are required to represent increasingly large numbers, the demand for improper representation also increases. However, the sense of propriety remains informal.

[9]Gottlob Frege, *The Foundations of Arithmetic: A logico-mathematical enquiry into the concept of number*, trans. J. L. Austin, 2nd revised ed. (Evanston, Northwestern, 1980). See esp. 6, 10, 11, 19, 34, 38, 69, 70, 93, 97, 98, 101, 102, 105, 114.

[10]idem, 6, 10-11.

the name."[11] Indeed, the *psychological* status of sufficiently large numbers is on a par with that of the infinite: "That we cannot form any idea of an infinite Number is of absolutely no importance; the same is equally true of finite Numbers." Frege's insistence secures the rights of a number such as $1000^{(1000^{1000})}$ to count as an object, "whose properties we can come to know, even though it is not intuitable."[12] The historical irony would be that, as Coffa observes, Frege's insistence on the logical rights of such large numbers leads, in Russell's hands, to an intensive preoccupation with the status of the infinite at the expense of any specific attention paid to our knowledge of large, finite numbers. More importantly still, according to Wittgenstein it supported a false *picture* of the infinite, in which we tend to think of it as "huge," since in the Frege/Russell approach large numbers are conceptually "on a par" with infinite number(s).[13]

While I find the parallel Coffa draws between Husserl and Russell provocative and illuminating, and while there is no denying that Husserl tended to focus on the nature of representation and Russell on the status of the proposition, I will ultimately take issue with some of the morals Coffa draws from this state of affairs. In particular, I suggest that the so-called anti-psychologism associated with Husserl's turn, on the one hand, and Russell's propositionalism, on the other, is a much more mixed affair than the overwhelmingly positive development Coffa would have us believe. Coffa's perspec-

---

[11]idem, 38.

[12]idem, 114.

[13]See, e.g., Ludwig Wittgenstein, *Wittgenstein's Lectures on the Foundations of Mathematics Cambridge, 1939*, ed. Cora Diamond (Ithaca: Cornell, 1976), 255. This passage indites a certain way of thinking about the infinite, and does not directly implicate the Frege/Russell tradition. But Wittgenstein turns in the next lecture to the Frege/Russell view of the relation of logic to arithmetic, and in particular to the status of propositions about large numbers. This more specific indictment is subtle, and proceeds along lines akin to the diagnosis of Cantor's continuum I discuss below, so I will not pursue it further here.

tive is predicated on the belief that the distinction between mental representations and mind-independent propositions is vital to overcoming the scourge of psychologism, and that this overcoming is one, if not in fact the main, achievement of the semantic tradition whose historical trajectory he traces out. This attitude is, of course, that of a card-carrying member of that (particular) semantic tradition which Coffa unabashedly promotes; but for those of us who would recognize, at the very least, other possible cogent philosophical attitudes, Coffa's particular brand of semanticism must defend itself, in particular, against the charge of covert Platonism.[14] As it turns out (or so at least I will claim), Coffa leaves himself open to such a charge precisely to the extent to which he misses a distinction internal to anti-psychologism which the status of the parafinite helps us to see. He stands, in this regard, in good company: on this issue, both Russell and the later, anti-psychologising Husserl held the same view as Coffa.

---

[14]Or else, of course, acknowledge a Platonic attitude which there is no textual evidence for attributing to Coffa, but which Coffa does himself recognize as governing Russell's philosophical strategy: "Acquaintance and a Platonistic ontology were Rusell's twin answers to the problem of the a priori. His philosophy of logic was the immediate consequence of this" (STKC, 127). Coffa goes on to state that "Wittgenstein explained that [Russell's] whole approach was ill-conceived" (ibid.), but it is not clear to me that he goes on to give a picture of a semantic philosophy in which this ill-conceivedness is entirely avoided. Regarding Wittgenstein, in particular, Coffa notes that "[i]t isn't easy to decide whether Wittgenstein should be included among the members of the semantic tradition or among its most ferocious enemies" (STKC, 141). Beyond a matter of Wittgenstein interpretation, I think this ultimately points to problems with the very idea of a historical semantic tradition as such. I return to Wittgenstein in the next chapter.

5.2 Questioning Finitude: van Dantzig and Isles

In order to unearth the difficulty with large numbers, I turn now to some later attempts to deal with the status of the parafinite. D. van Dantzig's "Is $10^{10^{10}}$ a Finite Number"[15] and David Isles' more recent, "What Evidence is There That $2^{\wedge}65536$ is a Natural Number?"[16] both ask whether the predicate 'finite' may be said to apply equally to small finite numbers and (what have traditionally been taken to be) large finite ones. In both these articles attention is restricted to the natural numbers, and so at the very least the number 1 will count as a small number, and presumably we will have some supply of small numbers succeeding it. Are there, on the other hand, numbers which are in some sense large, and if so, in what sense? Are there natural numbers which are so large as not to be finite, at least in the way that small numbers like 1, 2, and 3 are?

Van Dantzig is the first person of whom I am aware to suggest explicitly that numbers such as $10^{(10^{10})}$ may not be finite. Van Dantzig's efforts were apparently motivated by Gerrit Mannoury's assertion in 1931 that the distinction between the finite and the transfinite cannot be operationally defined, and he further tells us that his paper is "an effort to relieve the apparent contradiction ... which originally made this statement [by Mannoury] ununderstandable [sic] and unacceptable to most mathematicians, including, till some years ago, myself."[17] The contradiction referred to

---

[15]D. van Dantzig, "Is $10^{10^{10}}$ a Finite Number?," *Dialectica* **9** (1956) 273-7, reprinted in Richard L. Epstein and Walter A. Carnielli, *Computability: Computable Functions, Logic, and the Foundations of Mathematics*, 2nd edition (Belmont: Wadsworth/Thomson Learning), 260-3 (references are to this later reprinting).

[16]David Isles, "What Evidence is There That $2^{\wedge}65536$ is a Natural Number?," *Notre Dame Journal of Formal Logic* **33** 4 (1992) 465-80.

[17]van Dantzig, "Finite," 262; Gerritt Mannoury, *Woord en Gedachte*,

is that "it is impossible to construct natural numbers as large as $10^{10^{10}}$, but $10^{10^{10}}$ *is* a natural number."[18] Here the sense of construction at issue is what van Dantzig calls actual constructibility, and the actual construction van Dantzig is declaring impossible here is that of constructing (physically) a sequence of $10^{(10^{10})}$ strokes ('|') or (mentally) a sequence of $10^{(10^{10})}$ mental acts.[19] Van Dantzig, however, believes that "[t]he contradiction, however, is apparent only, as one has meanwhile unconsciously changed the meaning of the term 'natural number'."[20] The upshot of van Dantzig's position is that we must recognize that we typically use the term 'natural number' in a variety of different ways which must be distinguished rather than being lumped together, as they typically are.

Van Dantzig's proposal has some dramatic implications. Suppose, he says, we accept as constructed a given collection S_1 of successors of 1 (that is, some chain of natural numbers beginning with 1, all of which have been actually constructed). On the basis of this construction "[t]he definition by complete induction of the sums:

$$x + 0 = x, \quad x + y' = (x + y)',$$

then is applicable only inasfar [sic] as $x + y' \in S_1$"[21] (Here ' denotes the successor operation.) It is quite a different thing, on the other hand, to form *formal* sums $x + y$, where x and y are drawn from S_1. The difference lies in the fact that in the first case we can understand properties of the sums at issue (e.g., $x + y = y + x$) to have received their *proof* through the process of construction, whereas in the latter case the analogous equalities would need to

---

(Groningen, 1931), 55-58.

[18]van Dantzig, "Finite", 261.

[19]For van Dantzig the first, physical alternative corresponds to the definition of number given by Peano, Whitehead and Russell and Hilbert, the second to that given by Brouwer.

[20]van Dantzig, "Finite," 261.

[21]ibid.

be understood as *postulates*. Perhaps the central consequence of van Dantzig's position is to call into question the role traditionally assigned to the principle of induction. In particular, Poincaré's position is no longer tenable:

> Poincaré's statement that complete induction is the creative principle of mathematics can not be maintained. Such a principle—inasfar [sic] as the term is appropriate —is contained in the successive definitions of arithmetical operations, their formal extension outside the class of numbers hitherto obtained, and the formal maintenance of the arithmetical rules proved for those which belong to the first class S_1.[22]

Specifically, when we introduce new natural numbers (now in the second, formal sense) through the process of introducing the exponentiation operation, we must view this as an extension of the original set of numbers in terms of a formalism that respects the rules laid out for the use of the original natural numbers (now in the first sense). Van Dantzig then goes on to give an explanation of the use of symbols like ω, Cantor's first transfinite ordinal, or $\aleph_0$, Cantor's first transfinite cardinal, to mean " 'numbers surpassing everything I can ever obtain' but not as anything essentially different from the numbers he *can* obtain."[23] The result is that "[t]he difference between finite and infinite numbers is not an essential, but a gradual one,"[24] just as the difference between formalist and intuitionist foundations of mathematics becomes on this picture. Van Dantzig in fact anticipates the so-called ultra-intuitionism of Yessenin-Volpin[25] when he remarks that "[a]n intuitionist, if he were consistent, might *not* call $10^{10^{10}}$ a finite natural number. He

---

[22]idem, 262.

[23]ibid. The problem is to understand what the sense of 'essentially' is here.

[24]idem, 263.

[25]A. S. Yessenin-Volpin, "The ultra-intuitionistic criticism and the antitraditional program for foundations of mathematics," in A. Kino, J. Myhill, and R.

also would reject (most of) today's socalled [sic] intuitionistic mathematics as being too formal. But—luckily perhaps—he is not always so consistent."[26]

In line with these comments about the intuitionist it ultimately seems that the question whether $10^{(10^{10})}$ is a finite number is not one which will receive a univocal answer according to van Dantzig. In fact, given van Dantzig's assertion that the distinction between the finite and the infinite is a gradual affair, the question threatens to become just as meaningless as it would have previously been taken to be pointless by those for whom the answer would (continue to) be an obvious 'yes'. Perhaps, then, the greatest potential merit of van Dantzig's project, if it is accorded any degree of credence at all, is simply to have thrown new light on the *questionability* of the distinction between the finite and the infinite. In particular, there is no positive account of how numbers such as $10^{(10^{10})}$ *specifically* (i.e. numbers which are large enough) could be taken to be parafinite. Indeed, van Dantzig remarks at one point that in the interest of simplicity he will leave out of consideration the fact stressed already in 1909 by G. Mannoury that the identifiability and distinguishability even of the smallest numbers are not absolute.[27] Here van Dantzig hints at a mathematical fallibilism that one finds already in the writings of Mill and C. S. Peirce,[28] but it remains unclear to what extent this is a position he has any interest in adopting. What *is* clear, although it is a point which seems largely

---

E. Vesley, *Intuitionism and Proof Theory* (Amsterdam: North-Holland, 1970), 3-45.

[26]van Dantzig, "Finite," 263.

[27]ibid. Here van Dantzig refers to Mannoury's *Methodologisches und Philosophisches zur Elementarmathematik* (Haarlem: P. Visster, 1909), in particular 6-8.

[28]On mathematics as an experimental science see, e.g., C. S. Peirce, *Reasoning and the Logic of Things* (Cambridge: Harvard University Press, 1992), Lecture Four, "The First Rule of Logic," 165ff. On the dubitability of even the simplest mathematical truths (Peirce mentions as an example twice two is four) see e.g. MS. 955 in the Harvard Collection of Peirce's papers.

independent of questions about the distinction between the finite and the infinite, is that van Dantzig wants to stress a distinction between construction and formalism which leads in the direction of exploding any unified conception of the natural numbers as a homogeneous collection.

This concern for the heterogeneity of our treatment of what has traditionally been taken to be *the* homogeneous collection of natural numbers is pursued in significantly greater detail in the more recent work of David Isles, to which I now turn. I begin by discussing a paper by Isles entitled "Remarks on the Notion of Standard Non-Isomorphic Natural Number Series,"[29] which attempts to present a development of some of the ideas of Yessenin-Volpin, and then I turn to Isles' "What Evidence Is There That 2^{65536} is a Natural Number?," in which Isles extends the investigation opened by the paper of van Dantzig I have discussed above.[30]

In the first of these two papers Isles begins by remarking that his article will be an attempt to "begin an explanation (independent, to some extent, of [Yessenin-]Volpin's work) of where such a viewpoint [as Yessenin-Volpin's] might lead."[31] Specifically at issue is Yessenin-Volpin's position that "it is possible to work consistently with 'natural number series' of various lengths each of which may be closed under some but not all primitive recursive functions."[32] Although the existence of non-standard models of arithmetic had been known as early as the 1920's, Isles' position will reverse the commonly held view that these exotic non-standard models are in

---

[29]David Isles, "Remarks on the Notion of Standard Non-Isomorphic Natural Number Series," in *Constructive Mathematics: Proceedings, New Mexico, 1980*, Springer Lecture Notes in Mathematics vol. 873 (Berlin: Springer Verlag, 1981), 111-34; partially reprinted in Epstein and Carnielli, *Computability*, pp. 263-70 (references are to the original printing).

[30]I discuss Isles' program at greater length in *Diagnosing Contemporary Philosophy with the Matrix Movies*.

[31]Isles, "Remarks," 111.

[32]ibid.

some way *richer* than the familiar natural numbers by claiming, instead, that the relevant objects involved in a formalization of the natural numbers, which Isles calls natural number notation systems are "in general ... less rich and complex than 'the' so-called 'intuitive' series."[33]

Isles points out that, by virtue of the central role they occupy within the mathematical enterprise, the natural numbers play a different role from many other mathematical objects. Specifically, he thinks that the principle of induction is something which is accepted *on the basis of* our understanding of the natural numbers. One consequence of applying the principle of induction *to* the natural numbers is then to move from a collection of objects closed under one (particularly simple) recursive operation, namely succession, to a collection of objects closed under *all* primitive recursive operations.[34] This is radically distinct from, and hence to be contrasted with, the case of (infinite) ordinal numbers, where arguments or new existence assumptions are often required to guarantee the closure of a collection of ordinals.[35] Prima facie (which is to say, before van Dantzig and Isles!) we characterize the distinction between finite and infinite numbers in terms of the (in principle) constructibility of the former versus the formally symbolic status of the latter; what both van Dantzig and Isles encourage is an importation of this distinction between construction and formalism into the sphere of what was formerly taken to be constructible. In this way, the sense of the distinction between construction and formalism is re-opened. With this move a relativization of the notion of construction inevitably occurs which involves a relativatization of

---

[33]ibid.

[34]Since the technical details do not concern us here, let me simply characterize primitive recursive operations roughly as those which can be achieved by processes of simple iteration of a certain sort (which can be defined formally). For details see, for example, George Boolos and Richard Jeffrey, *Computability and logic* (Cambridge: Cambridge University Press, 1974).

[35]Isles, "Remarks," 112.

the distinction between the finite and the infinite. In Isles' case this relativization takes the form of a replacement of the finite/infinite distinction by a distinction between the "arrived" and the "in the future," to which Isles adds the cautionary note that "a notation which is arrived, i.e. finite, with respect to one series is in the future, i.e. infinite, with respect to another."[36] In terms of this temporal language we see the sense in which Isles inherits the tradition of Brouwer and Yessenin-Volpin.

Isles dramatically pinpoints the difficulty which lies behind attempts to define the natural numbers by examining what he claims "forms the basis for most mathematicians' belief in the uniqueness of the natural numbers."[37] This is what Isles calls the counting description of the natural numbers, which may be described in terms of the following three rules:

R1) Write down a stroke $|$;

R2) Given a set of strokes (call it X) write down $X\,|$.

R3) Now apply R1 once and then apply R2 again and again.[38]

Here we are given a set of instructions for constructing natural numbers in van Dantzig's sense. The problem, as Isles points out, lies with R3: in order to understand this rule we are already relying on a conception of a natural number series in the specification that we apply R2 again and again; the attempted definition is circular.[39]

---

[36]idem, 116. The temporal language connects Isles' exposition to the intuitionism of Brouwer and the ultraintuitionism of Yessenin-Volpin.

[37]idem, 113.

[38]ibid; here I have substituted the symbol '$|$' for '1' so as to bring Isles' notation into conformity with his later paper.

[39]As Isles notes, the circularity in definition was already recognized by van Heijenoort in an article on Gödel's Theorem in the *Encyclopedia of Philosophy* (New York: MacMillan, 1967), vol. 3, 356. In *Thirty Years of Foundational Studies: Lectures on the Development of Mathematical Logic and the Study of*

Our ability to fix the application of R3 is radically underdetermined: for example, Isles mentions in particular the understanding of R3 which would sanction applying R2 twice, that is, literally again and [once more] again.

As soon as the application of induction to the natural numbers is questioned, the possibility of accounting for a unitary sense of the natural numbers is radically undermined; it is this loss of unity (despite his retention of a principle of induction) which stands at the center of Isles' second article, to which I now turn. Here Isles modifies van Dantzig's approach by asking not whether 2^{65536} is a finite number, but rather *what evidence* we have for the claim that it is a *natural* number. It may seem that, since natural numbers have traditionally been taken to be finite,[40] this amounts to much the same thing. Isles' point, however, is not so much to suggest that 2^{65536} either is or is not a natural number, let alone a finite one, as it is to suggest that the burden of evidence should be shifted onto those who would claim that it *is* a natural number in some definite sense. Thus Isles' primary goal is to call into question the view of most mathematicians that it is obvious that 2^{65536} *is* a natural number.[41] He begins by pointing out that "there have been

The Foundations of Mathematics in 1930-1964 (New York: Barnes and Noble, 1966), Andrzej Mostowski discusses a potential criticism (which Mostowski says was noted not long ago) of Gödel's incompleteness result proposed by L. Rieger, namely that it relies on the unquestioned assumption of a 1-1 correspondence between integers and numerals, pp. 22-3. Mostowski does not supply a reference, but see Rieger's related article, "Sur le problème des nombres naturels," in *Infinitisitic Methods (Proceedings of the Symposium of the Foundations of Mathematics, Warsaw, 1959)* (Oxford: Pergamon Press, 1961), 225-33.

[40]It should be remembered, however, that there are commonly taken to be *non*-standard *models* of the natural numbers which are taken to include infinitely large numbers. One of the consequences of the approach broadly advocated by van Dantzig, Yessenin-Volpin and Isles is that this way of construing models of the natural numbers becomes problematic in multiple regards.

[41]Isles, "Evidence," 465.

objections to the claim that all such exponential expressions name a natural number, two of the best known being due to Paul Bernays and Edward Nelson."[42] The issue is not pressed by Bernays, but is developed at length by Nelson, who specifically declares that 2^{65536} is *not* a natural number, *only* a formal construct. To a nominalist, Nelson remarks, including himself among their number, "it is clear where to draw the line separating the real from the speculative: R_5, which is a system of 65536 objects, exists—but R_6, with its $2 \uparrow\uparrow 5$ [i.e. 2^{65536}] members, is only a formal construct."[43] However, unlike Nelson's own predicative approach which limits the application of the principle of induction, Isles (motivated by his reading of Yessenin-Volpin) suggests an alternative route whereby the principle of induction retains its full validity but quantifiers are bound to distinct collections which may even change at various points in the derivation. The motivation for this approach has to do with Isles' desire to specify what it is that distinguishes exponentiation, in particular, from the earlier operations of addition and multiplication in order to explain why we might want to speak of exponential notations specifically as distinct from the natural number notation system associated with addition and multiplication. Isles is taking a stab at trying to describe why the exponentiation operation constitutes the sort of barrier reef, as I've described it above, which separates tame mathematical notations from ones that take us out into the mathematical deep. Thus, Isles will apparently be willing to grant *any* numerical notation involving only succession, addition and multiplication a basic status, but will suggest that those who wish to go on to include numerical notations involving exponenti-

---

[42]ibid; Paul Bernays, "On Platonism in Mathematics," reprinted in *Philosophy of Mathematics*, 2nd ed., ed. P. Bennaceraf and H. Putnam (Cambridge: Cambridge University Press, 1983), 258-71; Edward Nelson, *Predicative Arithmetic* (Princeton: Princeton University Press, 1986).

[43]Nelson, *Predicative Arithmetic*, 97. Although he does not say so explicitly, it is presumably on the basis of this specific declaration by Nelson that Isles considers the example of 2^{65536} in his paper.

ation along with those previously mentioned should be required to give some argument for taking such notations to be natural numbers on a par with the others. That is, when we move beyond the barrier reef, the burden of proof shifts onto those who would claim they are still working with natural numbers in some unproblematic sense. What is it, for Isles, that distinguishes exponentiation?

In order to address this question, which leads directly to what I take to be the key point of Isles' article, Isles turns to some traditional proof-forms which could be used to argue for the closure of the natural numbers under exponentiation. Isles' approach will then be to try to understand what could possibly distinguish these types of proof from the analogous proof-forms which would be used to demonstrate the closure of the natural numbers under addition and multiplication. He first considers the following proof, which, following Isles, I will refer to as Example 1:

1. Assume that we have proved $(\forall n)(\forall m)(\exists p)[n \cdot m = p]$.

2. From the recursion equations for exponentiation we have $q^\wedge \mathbin{|} = q$ (i.e., $(\exists s)[q^\wedge \mathbin{|} = s]$) for any natural number q.

3. As the hypothesis of induction we assume $(\exists s)[q^\wedge r = s]$. Then $q^\wedge(r \mathbin{|}) = (q^\wedge r) \cdot q$. Using the formula in Step 1 (with $q^\wedge r$ substituted for n and q for m), we conclude that $(\exists s)[q^\wedge(r \mathbin{|}) = s]$.

4. By induction on r, then, it follows from Steps 2 and 3 that $(\forall q)(\forall r)(\exists s)[q^\wedge r = s]$.[44]

Isles points out that something peculiar happens if we substitute in large values for r in step 3. Suppose we let $r = 65535$ and $q = 2$. We then get the implication

$$[(\exists s)(2^\wedge 65535 = s) \rightarrow (\exists s)(2^\wedge 65536 = s)];$$

---

[44]Isles, "Evidence," 467.

but, Isles argues, the truth of the conclusion can only follow from the truth of the antecedent, and the antecedent *assumes* that 2^{65535} is a natural number.[45] What we are involved with here is a more sophisticated version of the problem Isles pointed out regarding "again and again" in his previous paper: by virtue of the way the argument relies on the principle of induction it effectively assumes what it sets out to prove.

However, as Isles goes on to concede, this recognition provides no direct help in understanding what could possibly distinguish exponentiation from the more basic recursive operations of addition and multiplication.[46] Here the key, instead, is to focus on that feature of induction which makes it objectionable to Nelson: its *impredicative* status. In a passage which Isles cites, Nelson remarks that:

> It is not correct to argue that induction only involves the numbers from 0 to n; the property of n being established may be a formula with bound variables that are thought of as ranging over all numbers. That is, the induction principles assumes that the natural number system is given.[47]

Unlike Nelson, Isles' response to this feature of induction is not to renounce the full strength of the principle of induction but, rather, to require that formulas with bound variables be freed by assigning a (constructive) range to each instance of a bound variable. Here, again, we are dealing with a "burden of proof" type commitment: Isles wants to shift the burden of proof onto those who would unproblematically identify the ranges of all bound variables: why should we assume in advance they are all equal? Until such a burden of proof can be discharged, we must identify each bound

---

[45]idem, 469.

[46]idem, 470.

[47]Nelson, *Predicative Arithemetic*, 1; cited by Isles, "Evidence," 467-8.

variable with its own "personal" range of variation. When this is done the proofs of the closure of addition and multiplication of constructed natural numbers become discernibly different from the proof of the closure of exponentiation of natural numbers in a way which motivates an argument that natural number notations involving exponentiation should be distinguished from those simply involving succession, addition and multiplication. The details are beyond the bounds of what can be presented here,[48] but the basic idea is that when derivations are indexed to variable ranges, the derivation of the closure of the natural numbers under exponentiation remains question-begging in a sense related to that described in Example 1 above whereas those provided for addition and multiplication do not.[49] More specifically, Isles shows how the arguments for the closure of addition and multiplication only assume the closure of the successor operation whereas the argument for the closure of exponentiation tacitly assumes the closure of succession, addition, multiplication *and crucially* exponentiation.

Isles' result is important because it offers a suggestion for how the largeness associated with exponentiation leads to a *different* sense of natural number which is *not* associated with any specific notion of physical or mental constructivity. This leads to the first slogan I would propose for understanding the parafinite, and since this slogan embodies the numerical characterization of the parafinite, I will call it the *principle of the numerical parafinite,*

---

[48]I give a detailed outline in "The Director's Cut," in *Diagnosing Contemporary Philosophy with the Matrix Movies*. There, in the "Theatrical Version," I also present an axiomatic argument due to Edward Nelson (what I call the quick and dirty argument) which makes a comparable point by exploiting the non-associativity of the exponential operation, i.e. that in general $a^{(b^c)} \neq (a^b)^c$. For Nelson's original argument, see "Warning Signs of a Possible Collapse of Contemporary Mathematics," in Michael Heller and W. Hugh Woodin, eds., *Infinity: New Research Frontiers* (Cambridge: Cambridge, 2011), 76-85.

[49]see Isles, "Evidence," 471-3.

or **PF 0**:

> **PF 0:** *Parafinite numbers are numbers that are large in the logical sense.*

In Isles' analysis the largeness associated with exponentiation is *logical* because it is associated with the logical structure of the proof of the closure of the exponentiation operation which he analyzes. I call the above a slogan (in the spirit of sloganeering in category theory)[50] rather than a definition because I want to leave vague what exactly would be included under the idea of 'large in the logical sense'. I label it "Principle Zero" because it is the root from which the other principles derive, and also because (like the void) it is difficult to make much of it without further specification.

In the standard approach to arithmetic (say the standard model of Peano arithmetic) there *are* no numbers which are large in the logical sense: the parafinite has been erased without a trace.[51] In other cases, such as Robinson's non-standard analysis, the only large numbers turn out to be infinite ones. But there are other approaches in which the notion of a logically large "parafinite" (non-infinite) number makes sense. Petr Vopěnka's Alternative Set Theory (to be discussed below) and Reeb's program of non-standard analysis (already mentioned above) provide two examples. Isles' example, in any case, gives us an instance of it, and enough of an orientation that several implications of the slogan may already be drawn out. Here we may say that the entire natural number notation system (with exponentiation) is parafinite relative to that involving only succession, addition and multiplication. This is not a matter of not *knowing* whether $2^5 = 32$, but literally of 2^5 and

---

[50]For an example of such sloganeering see, e.g., J. Lambek and P. J. Scott, *Introduction to higher order categorical logic* (Cambridge: Cambridge University Press, 1986), 4, 5, 6, 15, 18.

[51]Unless one identifies this trace metamathematically, in the existence of undecidability results, a "blind spot," etc.

32 being different *numbers*. This is related to the new conception of number being attached to a scale which I mentioned in the prologue. Here numbers are dependent on the scale of operations which they take as assumed. '2^5' is a number in which we do not *assume* the existence of the exponential operation but *identify* it explicitly. Since we are 'calling' on the exponential operation explicitly, what we *assume* in this case are whatever operations (like, say, addition and multiplication) we would appeal to in order to work up exponentiation. We are free to "identify" the two numbers '32' and '2^5' as *equivalent* should we care to, but this involves establishing the sense of the equivalence *relation* which such an identification involves.

Think, if you will, of writing a computer program to produce the exponential operation. We might (and probably would) construct exponentiation as repeated multiplication, multiplication as repeated addition, and addition as repeated succession (i.e. adding one). So the functions we are *assuming* in this case are: succession, addition and multiplication. On the other hand, as long as we write '32' without attaching any *explicit* scale, we should assume that the full range of functional operations (whatever we might take that to be) *is* assumed.

In this way we avoid the mathematical fallibilism advocated by Mill, Peirce and an entire tradition of mathematical empiricists (although such fallibilism could still be warranted on independent grounds) while retaining the ability to distinguish between different natural number series. Isles has given us a blueprint for handling natural numbers once we accept that there is no single unique series of natural numbers, and, as Isles puts it, once this step is taken the method for handling natural numbers which he advocates is almost forced.[52] This opens up the possibility of understanding certain natural number notation systems as parafinite relative to others; at the end of this chapter I will have something more to say

---

[52]Isles, "Evidence," 470.

about a weaker sense in which *specific* numbers may be parafinite relative to others (i.e. relatively parafinite), but the basic idea is straightforward: one number is parafinite relative to another when it outstrips the scale associated with the latter one.[53]

It is now clear why, in the traditional frame in which distinctions between natural number notation systems are collapsed, we cannot ultimately make any sense of the parafinite at all. Current philosophers and mathematicians would no doubt tell us this is no great surprise, since no one has ever seriously proposed this idea in the standard context. But, as I have argued above in the chapter on Galileo, a clear indication of the parafinite is already in evidence in Galileo's *Two New Sciences*, and as we have seen in the following chapter it was this very idea of the parafinite which was assimilated into the notion of an *indefinite* infinite in the work of Descartes and (especially) Leibniz. Eventually this historical development led to the Cantorian conception of the transfinite; one of the many stories I cannot include here is the story of the inadequacies associated with the conception of the Cantorian transfinite as *infinite*. But it is safe to say that this story alone would justify arguing that the problem of the status of the parafinite has a central role to play in understanding modern approaches to metaphysics. One need only think of possible-worlds theorists with their continuum-many worlds, all of which would be unthinkable without the orthodox Cantorian conception of the transfinite. As Shaughan Lavine has remarked: "[a]nything that can be done to reduce the apparent strangeness of infinite mathematical objects—or at least to understand the ways in which they are strange—will send subterranean tremors throughout philosophy."[54] However, what I am suggesting is *not* that so-called infinite mathematical objects are made any less strange by being recognized as parafinite, only that they are

---

[53]To be sure, this ultimately requires an account of what it means for a scale to be associated with a number.

[54]Lavine, *Understanding the Infinite*, 246.

less infinite![55] That this should nonetheless, and, in fact, *all the more* send subterranean tremors throughout philosophy is my thesis here. Since the retranslation of the Cantorian transfinite back into its more appropriate conception as parafinite is too large a story to tell (and it is still too early to tell it), much less assess its philosophical implications, I will begin to sketch a historical defense of this thesis by turning back, in the final section of this chapter, to modern concerns with the representability of large numbers and, specifically, to Hrusserll. Although there is ultimately a huge historical canvas to piece together, Coffa's passing remark may serve as a warrant for *beginning* with Hrusserll.

5.3   Methods of Using a Name

In a passage by Yessenin-Volpin which Isles cites, Yesennin-Volpin gives a description of what he will intend by formalization: "The notion of 'formalization' must now be enlarged: 'to formalize' the use of a notion means 'to expose a method of using its name'."[56] Unpacking, this passage yields the following picture: we have a

---

[55]This should not, however, be construed as a *finitist* project for several reasons, the most prominent of which is that the distinction between the finite and the infinite has (apparently) been relativized. Nonetheless, although the philosophical project associated with the introduction of the parafinite, which I will call *paraphysics*, is not any more finitist than it is infinitist, its situation between finitism and infinitism is nonetheless asymmetrical because there are certain potential traditional ontological commitments (the continuum, the absolute) which seem to lie beyond the bounds of paraphysics precisely *because* of their deeply infinitary status. This is to say that paraphysics cannot be viewed as a sort of liberal finitism since the idea of the parafinite is ultimately no more successful in accounting for the nature of the continuum or the absolute than any (traditionally) finitist perspective. In fact I believe that all so-called finitist positions are covertly parafinitary; but for the time being this matter must be bracketed. In the next chapter, we see Wittgenstein's "last ditch" effort to render a logically definite account of the mathematical continuum.

[56]Isles, "Remarks," 111.

notion; we have a name for the notion, to which latter the name refers (i.e. the notion is its denotation); we expose a method of using this name; this gives the (formal) meaning of the notion. Meaning is, very broadly construed, use. However, and perhaps ironically, the notion of use here is *formal*, and consequently even broader than the radical constructivism which in some sense makes it possible in the first place, giving rise to problems which are familiar from the work of Ludwig Wittgenstein, and which I will be discussing at length below.

Although this picture is open to criticism in any number of different regards, what this passage from Yessenin-Volpin encapsulates gives us a working model of what is at issue in the enterprises of Hrusserll circa 1900 as described in the passage from Coffa cited above. As such, this allows us to tie debates about the parafinite, as they emerge in the work of van Dantzig, Yessenin-Volpin and Isles, back to issues of canonical historical concern for the development of twentieth century philosophy. The distinction between formalization and construction which emerges in the work of van Dantzig and Isles (I leave aside the more complicated case of Yessenin-Volpin) in the context of issues concerning the status of large (and in particular exponentially large) numbers is, I would suggest, a radicalization of early 20th century attempts in the work of Hrusserll to do justice to the nature of numbers. These attempts were largely abandoned, in both cases, due to a perceived threat or innovation: in the case of Husserl, the specter of psychologism, and in the case of Russell the miraculous discovery of Peano (traces of which we already find appended as footnotes to the 1900 draft of the *Principles of Mathematics*). Because Husserl's work in the *Philosophy of Arithmetic* is more directly and more obviously tied to the issues discussed above I will turn to it first.

Husserl's account of the nature of arithmetic in the *Philosophy of Arithmetic* orients itself in terms of the observation that what we are able to grasp intuitively so far as number is concerned rarely exceeds the first dozen numbers. Arithmetic, consequently, is concerned with making up for this defect. Indeed, this is the fundamental purpose of arithmetic: *if there were no need to do so there would be no need for arithmetic.*[57] As Husserl puts it to particularly dramatic effect, God has no need to arithmetize.[58] All arithmetic per se relies on the use of Brentanian indirect representations, roughly the improper representations discussed above, and which Husserl also calls symbolic, in distinction to the direct representation of the first few numbers of which we are intuitively capable.[59] In terms of its *genesis*, it is this which makes arithmetic what it is. This is so much the case, Husserl goes on to assert, that if we cannot write and name addition then we can't think it anymore either.[60] The *power* of arithmetic is precisely that it is *not* constricted by the specific psychological limitations associated with direct representation.

Although the *power* of arithmetic representation is a result of its remaining (relatively) *unconstrained* by what we are capable of representing intuitively (i.e. *directly* psychologically), it is important to note nonetheless that if it weren't *for* these psychological limitations there would be no arithmetic: our psychological limitations establish the *domain* of the arithmetical (genetically) even if they do not contribute specifically to its character. What Isles' work, in particular, calls into question is whether there is not in fact an indirect logical reflection of these psychological limitations in the character of mathematics itself. Either way Husserl's account of arithmetic remains irrevocably linked to an appeal to psychology which he would later disavow. Indeed, as Coffa points out, Frege

---

[57]Edmund Husserl, *Philosophie der Arithmetik, mit ergänzenden texten (1890-1901)*, ed. Lothar Eley (The Hague: Martinus Nijhoff, 1970), 191.

[58]idem, 192n.1. Compare Hans Blumenberg, " 'Der Mensch zählt immer'," in *Beschreibung des Menschen*, ed. Manfred Sommer (Frankfurt: Suhrkamp, 2006), 318-77.

[59]idem, 190; see also STKC, 101 for a discussion.

[60]Husserl, *Philosophie der Arithmetik*, 187.

quickly criticized Husserl's *Philosopy of Arithmetic* for its psychologism (STKC, 102),[61] although Husserl only much later recanted, as he reports in the 1913 preface to the *Logische Untersuchungen*.[62] Husserl's enterprise was then to turn in the direction of an investigation of the givenness of the logical rather than in the direction of an appeal to a 'psychological origin' for the logical structure of any given phenomenological domain. In 1913 he remarked, looking back on his earlier work:

> But how symbolic thinking is possible, how the objective, mathematical, and logical relations constitute themselves in subjectivity, how the insight into this is to be understood, and how the mathematical in itself, as given in the medium of the psychical, could be valid, this all remained [in the earlier work] a mystery.[63]

The decision made by Husserl involves the hope that a non-psychological account of the possibility of symbolic thinking *could* be given, and conversely any competition or conflict between the psychological and the logical is renounced in the name of antipsychologism. The status of numbers as large is now seen as something *merely* psychological, whereas earlier *all* numbers participated in Number's largesse by virtue of the grounding of arithmetic in the attempt to overcome our inability to deal with large numbers intuitively.

By 1913 Husserl had definitively recognized the unsatisfactory status of his early account of arithmetic as a function of its attempt to derive the concept of number from the concept of collection: it

---

[61]Frege's criticism of Husserl dates from 1894; Husserl's *Philosopy of Arithmetic* was published in 1891.

[62]Husserl, *Introduction to the Logical Investigations,* trans. Philip J. Bossert and Curtis H. Peters (The Hague: Martinus Nijhoff, 1975), cited STKC, 102. But see also Eley's remark in his introduction to Husserl's *Philosophie der Arithmetik*, xxviii.

[63]Husserl, *Logical Investigations,* 35.

was this that inevitably committed Husserl to the psychologism he would eventually renounce.[64] Russell, on the other hand, was already circa 1900 so far from recognizing any reduction of number to collection or vice versa that his draft of the *Principles of Mathematics* begins with independent accounts of these two ideas. As Russell puts it, "the two ideas are on the same level of logical priority," and both are required in order to given an account of addition.[65] In this way Russell made it possible already in this early draft to provide an essentially logical account of these two notions, but at the expense of positing them as logically independent. He effectively distinguishes between what he calls pure numbers, which he takes to be indefinable (since, in particular, they cannot be defined by addition, for addition "is not a method of forming numbers, but of forming collections of which numbers can be asserted"),[66] and what might be called denotative numbers, which are numbers insofar as they are asserted of collections. It is with the latter that we are involved when we add, while it is on the basis of the former, along with the idea of collection, that Russell is able to provide an *account* of addition. On this view, addition turns out to be something quite removed from the essence of number, which finds its expression in (irreducibly extrinsic) *relations among* these indefinable pure numbers. Because, at least in the case of the integers, these relations are most naturally understood as ratios, pure numbers turn out, in fact, to have a closer connection to multiplication than they do to addition.[67] This is a direct consequence of

---

[64]idem, 34-5.

[65]Bertrand Russell, *Toward the Principles of Mathematics 1900-1902: The Collected Papers of Bertrand Russell, vol. 3* (London: Routledge, 1993), 15. To some extent Russell's account draws on that he had proposed in his 1897 article, "On the Relations of Number and Quantity," *Mind* n.s. **6** (July 1897): 326-41, reprinted in Bertrand Russell, *Philosophical Papers 1896-99: The Collected Writings of Bertrand Russell, vol. 2* (London: Unwin Hyman, 1990).

[66]Russell, *Toward the Principles*, 17.

[67]This brings the notion of pure number into close relation to that of a di-

Russell's view that addition is a method for forming collections, not numbers. In any case, the distinction between pure number and the operations of arithmetic is not seen as having anything to do with our limited capacity to represent numbers intuitively. Instead, Russell hypostatically imbeds the distinction within the logical realm as a distinction between pure and denotative number.

Russell, on this point already allied with Frege (to some extent),[68] beat Husserl to the anti-psychological bandwagon. Consequently, as Coffa makes clear, since the problem of large (finite) numbers is merely a psychological one, the (logical) problem of the representation of large numbers became for Russell largely a problem about the nature of the *infinite*.[69] In particular there is a problem with the account that Russell was inclined to give of number when applied to infinite collections: that we would need to countenance infinite conjunctions.[70] Similarly to Husserl's distinction between intuitive and symbolic numerical representations, Russell was relying on a distinction between pure number and denotative number or number by representation (my phrases), but the details were quite different. In any case, Russell turned in the direction of what he would later call schemes by focusing on the way in which 'any' functioned. As Coffa judges this development,

What Russell had come to see in 1901-2 was that per-

---

vided whole. There seems to be a competition between the ideas of whole/part and collection at the root of Russell's exposition. An analogous tension is found in Cantorian set theory in terms of the interplay between the element and containment relations.

[68] see STKC, 386n.5.

[69] This is implied by Coffa's remarks at STKC, 103; but see also my remarks below concerning Russell's preoccupation with substitutability for expressions referring to large numbers.

[70] Coffa remarks that Frege had already considered and abandoned this idea: see STKC, 386n.4, with reference back to 70 as well. This problem will haunt Wittgenstein also.

haps he had been wrong when he told Moore than *any number* cannot be a concept. Perhaps the reason we can deal with infinity in propositions of finite complexity is that quantifier expressions signify the presence of a very peculiar sort of concept, one that, unlike normal concepts, does not just sit there in the proposition, either linking its partners into a unified complex or allowing itself to be the focus of referential attention. Instead, these denoting concepts, as Russell called them, play an altruistic semantic role in that they somehow manage to refer to (or, as he sometimes puts it, indicate) other objects, thus allowing the proposition to be about things other than its constituents. (STKC, 104)[71]

These peculiar concepts point in the direction of a resemblance with Frege's account of quantification, in which, as Coffa puts it, quantifiers are understood as specific second-order concepts (STKC, 104). Russell's account of schemes was clearly in place by the time of the publication of the *Principles of Mathematics* (STKC, 105). Denoting concepts come to play a role previously filled by indirect or symbolic representations: A denoting concept, Russell explained, "need in no way resemble what it denotes" (STKC, 105). They are, so to speak, *utterly* improper. Hence, in particular, large/infinite numbers may be denoted by concepts of much smaller complexity. Is this, then, to say that the notation '2^{65536}' is conceptually speaking simpler than the number it denotes? And if so, how, exactly?

Such issues would lead directly to the heart of the development of a theory of denoting. Yet there is at least one indication (which Coffa does not mention) that the replacement of symbolic representation by a theory of denoting did not leave the problem of large (finite) numbers entirely behind. This example occurs in

---

[71]The notion of 'semantic altruism' is related to Leibniz's development of the syncategorematic character of the concept 'infinite', discussed above.

the context of Russell's concern about the substitutability of various definite descriptions within a proposition in a manuscript from 1903. His question is: given a subject S, of which many descriptions are possible, will all the propositions in which such descriptions are substituted for S be about S? And if so, what can be the meaning of *about*?[72] Here Russell gives the example:

> Suppose we say 'the number of people at the meeting was very great', and suppose the number was, as a matter of fact, 5432; can we be said to be making a statement about the number 5432? By the substitution of identicals our statement becomes '5432 was very great'—a foolish statement, in which the past tense is out of place. And we should never have made the judgment if we had known that the number was 5432; we should have said 'the number of people at the meeting was 5432'.[73]

Russell even extends his example in such a way that substitution leads to falsehood by appealing to the example 'The number of people at the meeting was greater than any one expected'.[74]

There are several aspects of Russell's treatment of this example that are immediately questionable, but to discuss them would lead astray.[75] The sole point needed here is that the truth of statements about numbers (and even what they are about) depends on the description of the number which is given. Post van Dantzig and Isles we could say that this is because the notion of a unified conception

---

[72]Russell, *Foundations of Logic*, 317 (from 1903).

[73]ibid.

[74]ibid., n. 3.

[75]Whitehead criticized what Russell had to say about this example; the criticisms are given at 356 of Russell, *Foundations of Logic*. These criticisms were apparently discussed in correspondence between the two, but I have not been able to locate this correspondence.

of number is a myth, but that sort of option certainly isn't one that would have made Russell, or for that matter Husserl, happy.[76]

In any case, Russell's treatment of addition leads directly to the distinction between meaning and denotation. The meanings '2' and '1+1' are different, but unless the *denotations* are one and the same there will be two different even primes. Horrors!![77] Therefore the denotations *must* be the same. That we find so much conceptual resistance demonstrates just how far we have come from the Greek conception of number, in which there were *any number* of different twos!

It is just this inference which is fundamentally cast into doubt by the papers of van Dantzig and Isles. I would add that we should see this as instancing a residual discomfort with the framework which has become *de rigueur* since the revolutions of Russell and Husserl shortly after each of the respective works discussed here. In the 1900 draft of the *Principles* Russell had already insisted that 1+1 not be used to *define* 2. This insistence has come back to haunt him, and us: it is unclear that the issue he attempted to address there has either been successfully handled *or* circumvented.

5.4 Conclusion (for this chapter and for Track A)

The status of the parafinite, at least as I would promote our thinking about it, is logical, but in such a way that my promotion of it is in particular lack of sympathy with early 20th century antipsychologism. That is, there is a way in which the psychological

---

[76]Edward Nelson, "Mathematical Mythologies," in Jean Michel-Salanskis and Hourya Sinaceur, eds., *Le Labyrinthe du Continu* (Paris: Springer, 1992), 155-67, and more generally, Hans Blumenberg, *Work on Myth*, trans. Robert Wallace (Cambridge: MIT Press, 1985).

[77]Russell, *Foundations of Logic*, 358.

limitation which is the origin of arithmetic for Husserl circa 1900 ultimately must itself find a logically imbedded expression in the formal (i.e. syntactic) calculus of addition. When we talk about a number being large, we should think of this as having to do with fixing the scale of a particular quantifier. This could be done in a variety of ways, which would constitute various formally specified alternatives for investigating the parafinite.[78] But there are also points which are independent of *particular* scale although they are dependent on the logical structure of scaling. Such, for example, is the point Isles makes about exponentiation. Each of these scaling-based issues is an integral aspect of the investigation of the parafinite. We may formulate this as our second slogan about the parafinite. Since this slogan concerns the characterization of the parafinite in terms of quantification, I will call it the *principle of parafinite quantification*, or **PF 1**:

PF 1: *The parafinite is the logically large as you please.*

We should begin by thinking of this in game-theoretic terms as a competition between two parties: however big a number you give me I can go bigger. The first number fixes a scale and the second number is parafinite relative to that scale. In the American community, the work of Jaakko Hintikka has made clear the deep dependence of quantification on the dialogical model of logic;[79] in Germany, this point was already explored extensively by Paul Lorenzen.[80] Hintikka's work goes beyond Lorenzen in pointing out that there are subtle but pervasive forms of quantifier dependence

---

[78]Because I do not view 'the parafinite' as a univocal concept, I fully embrace such a plurality of perspectives.

[79]Jaakko Hintikka, *The Principles of Mathematics Revisited* (Cambridge: Cambridge University Press, 1996), esp. Chapter 2. See also Wilfrid Hodges, *Building models by games* (Cambridge: Cambridge University Press, 1985).

[80]for a selection in English see Paul Lorenzen, *Constructive philosophy*, trans. Karl Richard Pavlovic (Amherst: University of Massachusetts Press, 1987).

that have not been recognized in the standard approach to quantificational logic, specifically in a nonlinearity (or noncommutativity) in quantifier dependence which Hintikka identifies.[81] What is at issue in Isles' (and related)[82] work is a different sort of quantifier dependence, but also gives rise to non-linear quantification relations.[83] It is too soon to tell how Hintikka's quantifier dependence specifically relates to that under discussion here, but in any case we are already in a position to give a preliminary account of the parafinite. It is that collection of issues which involves the overstepping of bounded quantificational domains. How the originators of various philosophical traditions chose to sidestep this issue (Frege, Russell, Husserl) provides key insight into the particular trajectory of their respective approaches. But in *all* these cases the failure to address the status of the parafinite *directly* led to a series of prejudicial attitudes toward psychology on the one hand and metaphysics on the other that dominated 20th century approaches to philosophy with near uniformity. This is a story which remains largely to be told: in order to evaluate the status of metaphysics in the new millenium we need first to turn to the problem of the status of the parafinite.[84]

---

[81]Hintikka, *Principles*, esp. Chapter 3.

[82]A game-theoretic approach to the issue of quantification is already clearly indicated in van Dantzig, "Finite," see 262. There are quite analogous issues at play in the work of Jan Mycielski and (taking Mycielski's work as a point of departure) Shaughan Lavine. See in this regard, Mycielski, "Analysis Without Actual Infinity," *Journal of Symbolic Logic* **46** (1981) 625-33 and Lavine, *Understanding the Infinite*, esp. Chap. 8.

[83]See David Isles, "Theorems of Peano Arithmetic are Buridan-Volpin Recursively Satisfiable," *Reports on Mathematical Logic* **31** (1997) 57-74, particularly the comments on 58. I discuss the Buridan-Volpin structures at issue in this paper at some length in the book manuscript *Diagnosing Contemporary Philosophy with the Matrix Movies*, referred to above.

[84]This first draft of this essay was written during my tenure as a Fellow of the Center for the Humanities at the University of Georgia, and was also the written version of a talk I gave through the Center for the Humanities, 6

September 2000. I would like to thank the audience for their active interest and stimulating questions and comments, and I would also like to thank Matthew Traut for many helpful conversations concerning these issues and René Jagnow for comments on an earlier version of this chapter.

Chapter 6

The Continuum According to Wittgenstein

(Our problems are not abstract, but perhaps the most concrete that there are.)

(TLP, 5.5563)

6.1 Introduction

In what follows I am concerned to articulate the way in which Wittgenstein's thoughts about the mathematical continuum, and in particular Cantor's consideration of it, become a point about which the changes in Wittgenstein's philosophical work crystallize during the period of his return to Cambridge and to concerted philosophical work in 1929 and 1930. As will become clear, this line of research overlaps in significant ways with understanding Wittgenstein's rejection of a phenomenological language, what we might gloss as a "fundamental language of description," and should aid in the assessment of the consequences of this rejection for his philosophical development. In this chapter, I lay some of the ground for this larger concern by showing ways in which Wittgenstein's thoughts about the mathematical continuum presage certain of the points he will make about the notion of a phenomenological language. In particular, I suggest that Wittgenstein's rejection of Cantor's conception of the mathematical continuum serves as a model for his later rejection of a phenomenological language. In this way, I believe what I will present here opens up a powerful, and largely novel, perspective for assessing Wittgenstein's later philosophical work. In conclusion, I will suggest some ways in which there are problems opened during this period of Wittgenstein's work which motivate the direction of his later thinking up through, and including, his latest writings in the collection *On Certainty*.[1]   In particular, the perspective I suggest has the attractive feature that it offers a new suggestion for what Wittgenstein might have felt was new and prospective about lines of thought opened up in this final phase of his philosophical career.

These are consequences for our understanding of Wittgenstein's philosophy, one of the major philosophical endeavors of the twentieth century, and for his thoughts on the foundations of mathematics in particular, which have received proportionately less attention than any other major sector of his work. But this evaluation fits into the project of this book because it represents a major attempt to rethink our commitment to the notion of mathematical continuum. The articulation of the nature of the mathematical continuum and of the concept of number are the two foci around which our elliptical considerations of the parafinite have orbited along the course of this volume. In general terms my thesis will be that Wittgenstein's remarks provide powerful intimations of the parafinite which are, however, fundamentally limited by his retention of a conception of logical determinacy. What this means will

---

[1] Wittgenstein, Ludwig. *On Certainty*, ed. G. E. M. Anscombe and G. H. von Wright, trans. Denis Paul and G. E. M. Anscombe (New York: Harper & Row, 1969).

154

only become fully clear in the course of investigation, but another vantage on the same point may help to fill out the picture provisionally. We could also say that Wittgenstein's philosophy shows us that the "problem of the parafinite" is not one that can be addressed only by looking at the grammar of mathematics: resolving the problem at this level only shunts it back onto the question of logical form. It is at this level of logical form that Wittgenstein resolutely refuses to face the problem of the parafinite. But even in this regard his work represents a great resource for the "paraphysician," for it gives us an extensive and lucid picture of what happens when this problem is driven into the logical domain.

In another direction, I will suggest some ways in which Wittgenstein's approach to issues in the foundations of mathematics raises issues about an ethical perspective on mathematical praxis which may be of use for larger concerns in the foundations of mathematics and beyond. Indeed, the work I am conducting here is designed to be part of a larger framework for assessing the ways in which decisions made in the foundations of mathematics– regarding, for example, ontological commitments, the way in which the distinction between the finite and the infinite is drawn, and the emphasis placed on formalism– orient larger issues about the way in which mathematical research is conducted. I am ultimately concerned about this not just as an issue in the philosophy of mathematics, but as a metaphilosophical issue as well, since historically the foundations of mathematics have repeatedly served as a proving ground for metaphysical concerns. Wittgenstein's stance with respect to foundations of mathematics, in particular, is indicative of a broader attitude toward the task of philosophizing in general, and it has largely been the result of an incapacity to see Wittgenstein's remarks about the foundations of mathematics as a valuable and coherent part of his larger philosophical project that has prevented Wittgenstein's work in this area from receiving the attention I believe it deserves, and which it has recently begun to receive. I would hope that the work I initiate here would help to redress this imbalance in the reception of Wittgenstein's work, and thus would serve as an example of the way in which considerations in the foundations of mathematics can be an integral, and indeed even central, part of broader philosophical investigations. It also serves, near the end of this book, to indicate the prospect for a transition from concerns about the *parafinite* to a general philosophical reorientation drawing inspiration from the rethinking of the distinction between the finite and the infinite that the parafinite indicates. This general philosophical reorientation is what I call *paraphysics*, a program for philosophical work which I elaborate elsewhere.[2]

6.2 Background

6.2.1 Königsberg September 1930

From the 5th to the 7th of September 1930 in Kant's hometown of Königsberg, the Vienna Circle organized a symposium on the foundations of mathematics at which expressions of the logicist, intuitionist and formalist positions in the foundations of mathematics were presented. Rudolf Carnap presented the case for logicism, Arend Heyting for intuitionism, and John von Neumann for formalism. All three of these contributions were printed in the second volume of the journal *Erkenntnis* under the banner, "Report of the 2nd Conference for Epistemology of the Exact Sciences in Königsberg 1930."[3] Three contributions to the conference were not included in the volume. Two of these were not included because

---

[2]The two book-length manuscripts in which I try to motivate this program are *Kant, Shelley and the Visionary Critique of Metaphysics*, in which I develop an approach to philosophy called paraphysics, and *Diagnosing Contemporary Philosophy with the Matrix Movies*, which inscribes paraphysics within a larger project I call matrix philosophy.

[3]"Bericht über die 2. Tagung für Erkenntnislehre der exakten Wissenschaften Königsberg 1930," *Erkenntnis* **2** (1931), 87-190.

of their highly technical mathematical nature, but, it was hoped, would soon be published elsewhere.

The third contribution derived from a lecture for which the associated paper was not completed; this, it was hoped, would appear in a later volume of *Erkenntnis*.

Of the two technical contributions to the conference, one was a twenty minute presentation by the 24 year old Kurt Gödel of results from his dissertation, which appeared in article form in the 1930 *Monatshefte für Mathematische Physik* under the title, "The Completeness of the Axioms of the Functional Calculus of Logic."[4] Much more significant than this presentation, in retrospect, was a remark which Gödel made during the 7th September discussion on the foundations of mathematics, an edited version of which appears in the *Erkenntnis* volume. In its printed form, Gödel's brief remark runs as follows:

> One can (assuming that classical mathematics is consistent) furnish examples of propositions (and, indeed, of the form of Goldbach's or Fermat's), which are indeed materially correct, but which are unprovable in the formal system of classical mathematics. If one adds the negation of such a proposition to the axioms of classical mathematics, one obtains a consistent system in which a materially false proposition is provable.[5]

---

[4] "Die Vollständigkeit der Axiome des logischen Funktionenkalküls," *Monatshefte für Mathematik und Physik* **37** (1930) 349-360, reprinted with face-à-face translation by Stefan Bauer-Mengelberg in Kurt Gödel, *Collected Works*, vol. 1: *Publications 1929-1936*, ed. Feferman et al. (New York: Oxford Press, 1986), 102-123.

[5] "Man kann (unter Voraussetzung der Widerspruchsfreiheit der klassischen Mathematik) sogar Beispiele für Sätze (und zwar solche von der Art des *Goldbach*schen oder *Fermat*schen) angeben, die zwar inhaltlich richtig, aber im formalen System der klassischen Mathematik unbeweisbar sind. Fügt man daher die Negation eines solchen Satzes zu den Axiomen der klassischen Mathematik hinzu, so erhält man ein widerspruchsfreies System, in dem ein inhaltlich

As John Dawson has noted, no response to this remark is to be found in the rest of the discussion transcript, and in fact Dawson makes a case that Carnap, at least, was quite some time in understanding Gödel's ideas.[6] On the other hand, Dawson reports that after the 7th September session von Neumann drew Gödel aside and pressed him for more information, and that two and a half months later, on the 20th of November, von Neumann was able to write Gödel announcing a corollary to Gödel's result.[7]

In historical retrospect, Gödel's announcement overshadows, one might even say overshatters, the significance of the conference as a canonical expression of the three trends in the foundations of mathematics, all three of which Gödel's result would drastically affect. So much more easily, then, has history been able to forget that, in fact, as Mathieu Marion has put it,

> ... the three musketeers were four, since Waismann also presented in this symposium a paper entitled 'Über das Wesen der Mathematik; Der Standpunkt Wittgensteins' ['On the Essence of Mathematics; Wittgenstein's Standpoint'], which he never submitted for publication.[8] (WFFM, 37)

This, then, is the third contribution to the conference which was not published in the *Erkenntnis* volume.

Marion gives three prime reasons for the neglect of this paper and the work it embodied, which are cumulatively conclusive:

---

falscher Satz beweisbar ist." *Erkenntnis* **2** (1931) 148. (So far as I am able to recall, the above is my translation.)

[6]John Dawson, "The Reception of Gödel's Incompleteness Theorem," in *Gödel's Theorem in Focus*, ed. S. G. Shanker (New York: Routledge, 1988), 74-95; here, 76-7.

[7]idem, 77-8.

[8]Mathieu Marion, *Wittgenstein, Finitism, and the Foundations of Mathematics* (Oxford: Clarendon Press, 1998), 37; hereafter cited in text as above.

... first, it was never really developed by either Wittgenstein or Waismann (or anyone else for that matter), and, secondly, as a result no one really understood how it differed from other standpoints (the confusion with logicism was frequent). Thirdly, no one at the time was able to see the affinities between Wittgenstein's ideas and Church's lambda-calculus, which appeared in 1932. (WFFM, 37)

Marion redresses these circumstances, arguing in particular for the third of these claims. But before turning to Waismann's paper, I will first turn to the *Tractatus*, and to the description of mathematics that is presented there.

6.2.2 Intimations of the Parafinite

In order to understand how Wittgenstein's work intimates a conception of the parafinite, we need to look first at how he understood the distinction between the finite and the infinite. Wittgenstein is indirectly committed to the infinite in the *Tractatus* by virtue of the way he understands quantifiers. His way to understand them is to eliminate them, replacing a universal quantifier by a conjunction of primitives. As Marion points out, in Frege's treament of quantifiers,

[w]hen the domain of quantification is finite (and surveyable) then the truth-value of the quantified statement could in principle be determined as a finite product, but when the domain of quantification is infinite (or finite but unsurveyable) this assumption would be open to question, because it would be claimed that we cannot determine its truth conclusively. (WFFM, 30-1)

Frege's fix would be to declare of the infinite totality of objects that, as Dummett puts it, "we, as it were, survey it in thought as a whole" (quoted, WFFM, 31).

Wittgenstein's approach would differ. As Russell described it, Wittgenstein understood generality to consist "only in specifying the set of propositions concerned, and when this has been done the building up of truth-functions proceeds exactly as it would in the case of a finite number of enumerated arguments p, q, r, ..." (quoted, WFFM, 32). Quantification involves an appeal to the infinite, then, to the extent that we may specify infinite sets (should such exist) to quantify over. Michael Wrigley, and Marion following him, believe that Wittgenstein was perfectly happy with this in the context of the *Tractatus*.[9] Marion also reports a remark Wittgenstein made to Kreisel (no citation is given) that in the *Tractatus* "he had put down his system in a finite setting without bothering about the infinite case, assuming that if a problem was to be found in the infinite case, then there would have already been a problem in the finite case" (WFFM, 34).[10] On the other hand, Wittgenstein would later see his neglect of the finite/infinite distinction as a crucial failing of the *Tractatus* (WFFM, 35). In fact, there are apparently two main sources of unrest that led to Wittgenstein's departure from the position of the *Tractatus*: the neglect of the finite/infinite distinction and an issue about propositions expressing degrees of quality; this latter Wittgenstein would treat, in particular, in terms of propositions about color (ibid). According to Marion, Wittgenstein's response was to react negatively to the introduction of infinite truth-functions proposed by his close friend Ramsey and move closer to a finitist standpoint (WFFM, 47).

Contra Marion, I think it is a mistake to think that the way Wittgenstein resolved the problem he identified with the finite/infinite distinction in the context of the *Tractatus* was by opting for the finite and against the infinite. Marion's mistake in

---

[9]I rely on Marion's resumé of Wrigley's manuscript, and omit the details of Wrigley and Marion's positions here.

[10]This neutrality on Wittgenstein's part is significant on its own terms; I suggest that it is a tacit anticipation of Wittgenstein's handling of the parafinite, about which I will have more to say later in the chapter.

this regard is an example of the larger problem associated with attempts to characterize finitist positions in the foundations of mathematics, and among other things it is the impasse to which this leads which the development of a concept of the parafinite seeks to surmount. Having made this programmatic claim, I am ready now to return to Waismann's presentation of Wittgenstein's position in his Königsberg paper.

6.2.3 Footnotes to Waismann in Königsberg

Waismann ties his presentation of Wittgenstein's views back directly to those presented in the *Tractatus*: the commitment to logic consisting of tautologies is retained, as is the principle that "what a symbol shows cannot be said by means of that symbol."[11] To this Waismann appends two theses pertaining to the proper "method—a manner of thinking—with which to look at related questions in the foundations of mathematics."[12] The two elements of the method are given in terms of the thesis that "in order to ascertain the meaning (*Bedeutung*) of a mathematical concept, one must pay attention to the use that is made of it, and, secondly, that in order to visualize the significance (*Sinn*) of a mathematical proposition one must make clear how it is verified."[13] Here the Fregean distinction between *Sinn* and *Bedeutung* is reconfigured in terms of verification and use, respectively: at least on Waismann's presentation there is no segregating Wittgenstein from the operationalism and verificationism of the Vienna School, at this point at

---

[11]Friedrich Waismann, "The Nature of Mathematics: Wittgenstein's Standpoint," trans. S. G. Shanker, in *Ludwig Wittgenstein: Critical Assessments*, Volume Three: *From the* **Tractatus** *to* **Remarks on the Foundations of Mathematics**, 60-67; here 60. The original can be found in Friedrich Waismann, *Lectures on the Philosophy of Mathematics*, ed. Wolfgang Grassl (Amsterdam: Rodopi, 1982), 157-67.

[12]idem, 61.

[13]ibid.

which he stands perhaps in closest proximity to this group. Waismann deletes several passages from the original text which signify the potentially most reductive, hence radical, consequence of this orientation: in particular, that "a proposition and a proof must be regarded and treated as one and the same thing in mathematics."[14]

Following directly upon the deleted passage from which the preceding quotation is drawn, there appears an inserted paragraph which is, in turn, deleted! It is upon this paragraph that I wish to focus here, and so I quote it in full. Waismann writes:

> That is the general logical attitude from which I will proceed. But we have to consider yet another point of view. Namely, it appears that these logical considerations touch deeply on the nature of the infinite. In fact the following explanations aim at nothing less than the definitive elimination of the infinite from mathematics, given that the infinite has been interpreted as a totality; and to remove the transfinite means of inference which are connected with it. What I would like to show is how this goal is to be achieved on the basis of Wittgensteinian logic. But I will only indicate the solution of this task for elementary arithmetic, where all of the essential thoughts already emerge. For the construction of analysis there exist attempts which are not yet developed well enough for me to want to speak about them.[15]

In what sense is this "yet another point of view"? Waismann would go on to discuss problems with the mathematical infinite, and in particular with the application of the mathematical infinite to the

---

[14]idem, 67n.4.

[15]ibid. It is perhaps helpful here to compare Jan Mycielski's program in "Analysis Without Actual Infinity," *Journal of Symbolic Logic* **46** (1981) 625-633. I return to Mycielski's program in the Envoi.

actual world, in such a way that he would explicitly part company with Wittgenstein. In particular, Waismann would later enlist a notion of the potential infinite which he could appeal to in order to avoid the problems concerning the continuation of mathematical series which Wittgenstein would consider in his discussion of rule-following.[16] In the 1930 lecture the focus, instead, is on the ostensibly logical distinction between a totality and a system: "[a]n empirical totality goes back to a property (a propositional function); a system to an operation." Waismann conceives of an operation as what we might call a differential of form: "An operation is the transition from one sentence form to another. An operation is what has to be done to the one form in order to turn it into another one."[17] But this notion of transition or transformation remains largely unanalysed in the context of Waismann's presentation, and it will lead both to Wittgenstein's preoccupation with rule-following, on the one hand, and, on the other, to Waismann's desire to identify an insight which will allow for the continuation of mathematical series.[18]

One might conjecture, then, that the reason for deleting the passage drawing attention to the connection between logical issues and the status of the infinite lay in a lack of clarity in the relation between the logical and (anti-)infinitary perspectives. And yet Waismann does indeed indicate the nature of his program with

---

[16]In "Infinity and the Actual World," dated January 1959; see Waismann, *Lectures*, 119. Marion also interprets Wittgenstein's position in terms of a commitment to the Aristotelian notion of the potential infinite, but I argue below that this is inaccurate. In any case, in the lecture "Infinity and the Actual World," Waismann explicitly recognizes the treatment he gives of the foundations of mathematics, and of the status of the infinite in particular, as his own, and as differing from that given by Wittgenstein.

[17]Waismann, "Wittgenstein's Standpoint," 64-5.

[18]Waismann, *Lectures*, 119. Wittgenstein refers frequently to insight in the *Philosophical Remarks*, but at one point in the typescript in a marginal comment added later he remarks, "Act of *decision*, not *insight*" (PR, 171). I will have occasion to return to this comment below.

respect to elementary arithmetic in the lecture. This program is motivated, first, by criticisms of the formalist and intuitionist programs, and in both cases the criticisms hang on issues of referential definiteness. Waismann points out, first, that on the formalist view, "the Peano axioms of arithmetic could just as well be satisfied if one takes the numbers 100, 101, … instead of the series of natural numbers."[19] Thus, propositions in which we use numbers to refer to objects (presumably whether they be empirical or otherwise; Waismann's specific example is to the five Platonic solids) are referentially indefinite, since on the formalist view the Peano axioms have many equally satisfactory models. This criticism may be question-begging, but the point here is to understand how Waismann attempts to motivate Wittgenstein's position. In this regard the point of the criticism is clear: the formalist does not give us a way to ascribe a meaning (*Bedeutung*) to any given number. An account of meaning is given by Russell, but an unsatisfactory one, for here the *Bedeutung* of mathematical terms depends in an unacceptable way on empirical facts about the world:

> Suppose, Russell says, that there were exactly nine individuals in the world. Then we could construct the cardinal numbers from 0-9, but 10, which is defined as $9 + 1$, would be the null class. Consequently, 10 – and all of the following natural numbers – would be identical with one another; viz. they would all be 0.[20]

The distinction between a totality, which depends for its constitution on the properties of the actual world, and a system, in which what is meaningful is a function of rules of syntax, is designed to overcome precisely this problem, and so render the meaning of numerical terms definite without any appeal to the actual world. In this way, Waismann says, "Wittgenstein's point of view is the con-

---

[19]Waismann, "Wittgenstein's Standpoint," 62.
[20]ibid.

sequence of thinking through the Russellian interpretation to the end, where it has been purified of the remains of a false empiricism."[21]

As we have seen, it is on just this point that Wittgenstein was to admit that the program of the *Tractatus* faltered: it was not sufficiently detached from the nature of actual reality, particularly regarding Wittgenstein's agnosticism with regard to the finite/infinite distinction during this period. Hence Waismann is unjustified in his claim that the elimination of residual empiricism could only be achieved by descending deep down to the fundamentals of logic. It is just this pioneering work which is accomplished in the *Tractatus Logico-Philosophicus*.[22] The work of the *Tractatus* provided a start, but the report on Wittgenstein's standpoint in Waismann's lecture remains *in medias res*.

It is the goal of what follows to retrace the trajectory which leads from Wittgenstein's return to philosophy in the late 20's to the time at which Waismann presents Wittgenstein's position at the Königsberg Congress. The path itself is arduous, and failing an exhaustive investigation of Wittgenstein's notebooks from this period (which will not be attempted here) a mixed strategy is required, much like the tacking back and forth which is often required to traverse difficult terrain. My strategy for this traversal may be outlined roughly as follows: first to look at the reaction to the earlier work in the *Tractatus* as it is presented in two important pieces of secondary literature, then to jump forward to a key section of the *Philosophical Remarks*, upon which I will comment in detail. Then, with the perspective gained from this exercise, I offer my own criticisms of David Stern's and, more crucially, of Mathieu Marion's reconstruction of this trajectory, relying heavily at this point on further evidence provided by Wittgenstein's notebook material. Marion's interpretation of the development of

---

[21]idem, 66.
[22]ibid.

Wittgenstein's thought during this period provides a powerful (and worthy) opponent for the sort of interpretation I will provide, and it is first on the basis of my capacity to meet the implicit challenge posed by Marion's stance that my reconstruction must be judged.

There remains one final note needed in preparation for the trek: what follows is a much more detailed analysis of "conditions on the ground" than anything I've presented previously in this volume. This is necessary for two reasons: first, because this trajectory has not, in my opinion, been sufficiently appreciated, and second, because it involves the analysis of a deep but concomitantly often subterreanean shift in the major philosophical landscape which is Wittgenstein's ongoing thought. I think the payoff is worth the effort: we get to see, as "firsthand" as is historically possible, a major thinker working through a fundamental transformation in his thinking. Moreover, this transformation bears focally and powerfully on the theme of this volume: how the parafinite intimates itself at the level of the some of the deepest and most sustained philosophical reflections our intellectual tradition has yet witnessed.

6.3   Wittgenstein in Transition: Toward the *Philosophical Remarks*

Following on earlier discussions with the Vienna Circle, Wittgenstein's preoccupations upon returning to Cambridge in 1929 circled around problems in the foundations of mathematics, on the one hand, and, on the other, with the prospect of a physical language of phenomenology. In the latter regard, members of the Vienna Circle drew inspiration from the *Immanenzphilosophie* (Philosophy of Immanence) of Ernst Mach (see WA, 1, vii).[23] Both David Stern and

---

[23]Ludwig Wittgenstein, *Wiener Ausgabe: Studien Texte*, vols. 1-5, ed. Michael Nedo (repr. Frankfurth am Main: Zweitausendeins, 1994-1996); here vol. 1 xii; hereafter cited as above in text.

Mathieu Marion have discussed Wittgenstein's transitional writings from 1929 in recent book-length works, but in neither case is there an entirely focal recognition of what binds these two sets of issues together. At least at the most basic level the connection is obvious, and so it is perhaps this fact, along with the difficulty of articulating anything insightful about this obvious connection, which leads both these authors to neglect it; part of the blame, further, may lie with Wittgenstein himself. In any case the connection between Wittgenstein's concern with the philosophy of mathematics, circling particularly around questions regarding the mathematical infinite, and his concern with a phenomenological language intersect with respect to the problem of the way that the philosophical architecture outlined in the *Tractatus Logico-Philosophicus* depended on the givenness of the world of facts. As we have seen above, during the composition of the *Tractatus* Wittgenstein remained blissfully agnostic about the possibility of an empirically given infinity, and he also failed to recognize the full consequences of the fact that the mathematical infinite could not, in any case, be made to depend on the empirical state of reality. Further, in the *Tractatus*, Wittgenstein had taken infinity to be a number: he told his student John King in 1930-1 that in the *Tractatus* he had made this mistake, in particular, in supposing that there can be an infinite number of propositions (cited, WFFM 183).[24] He was later to admit that "in the *Tractatus* [he] had made the mistake of supposing that an infinite series was a logical product– that it could be enumerated, though we were unable to enumerate it," and to Georg Henrik von Wright he acknowledged in 1939 that "the biggest mistake [I] had made in the *Tractatus* was that [I] had identified general propositions with infinite conjunctions or disjunctions of singular propositions" (both cited, WFFM, 85; the first is

---

[24]see also STKC, 405n.5, which cites Ludwig Wittgenstein, *Wittgenstein's Lectures, Cambridge, 1930-1932*, ed. D. Lee (Totowa, N. J.: Rowman and Littlefield, 1980).

from lecture notes of G. E. Moore).

Wittgenstein had finessed the problem of the nature of atomic facts in the *Tractatus* from the beginning, but there are some general positive and a few more specific negative things to be said. On the first count: the recourse to atomic facts was from the beginning in the service of logical definiteness: all logical form was to be driven back to the logical form of elementary propositions, and these would mirror the forms of atomic facts. On the negative side, Wittgenstein believed that the particular structure of these atomic facts would be a matter of scientific, *not* logical-philosophical, investigation, and that the task would be a difficult one. Even more specifically, Wittgenstein reacted negatively to Carnap's restriction to two-place relations in describing these forms (see WFFM, 115 and 143). But he remained at once convinced that these forms could be empirically determined and that their specific nature remained irrelevant to the project of the *Tractatus per se*.

By the time that Wittgenstein submitted the typescript which was to become *Philosophical Remarks* to Russell in 1931, this situation was to have changed drastically. Not only had Wittgenstein abandoned the commitment to atomic facts and corresponding atomic propositions being empirical and hence philosophically irrelevant, he had indeed abandoned the idea that such an elementary language of atomic propositions was even a coherent one. Furthermore, by this point he had investigated, and then abandoned, the idea that this primitive stratum of independent elementary propositions could be replaced by a phenomenological language which occupied an analogous elementary position but did not consist of independent primary propositions. Indeed, he was convinced that language needs no such support in a primitive propositional stratum. In what follows I prepare to retrace this sea-change by considering Wittgenstein's rejection of the Cantorian conception of the continuum, but first I turn to the larger concerns of the *Philosophical Remarks* to situate these concerns.

6.3.1 *Philosophical Remarks*: Outline of a Position

In this subsection I outline part of Wittgenstein's position in the *Philosophical Remarks*, which must itself be viewed as an outline: Wittgenstein presented the typescript to Russell, which in revision was to become *Philosophical Remarks*, as an outline of his research since returning to Cambridge in 1929. Since Wittgenstein himself compiled this typescript we may assume that it represents a snapshot of his position at the time of compilation. Since he assembled it in haste, we may also assume that it is a sketch of his position rather than a full-fledged portrait. For this reason, my strategy will be to test the picture we receive here in some specific cases against the manuscript evidence provided in the definitive edition of Wittgenstein's works, the *Wiener Ausgabe*. My presentation of the *Philosophical Remarks* is already tacitly informed by this evidence, so in this regard the game is trumped, but turning later to the manuscripts will allow me to fill out the picture I provide of the *Remarks* and cast it in a more explicitly developmental light. Starting with a discussion of the typescript, on the other hand, has the dual merits of beginning with the source which is more widely received and also with a somewhat more systematic presentation.

A primary source of confusion regarding the material which Wittgenstein presents in the *Philosophical Remarks* stems from attempting to find in the remarks concerning the grammar of color language an argument for the rejection of the search for a primary, or phenomenological language. Wittgenstein asserts that he no longer takes the latter as a goal, clearly with reference to the commitment to the grounding of the analysis of language in an appeal to elementary propositions in the *Tractatus*, and as still exemplified, albeit in a modified form, in the essay "Some Remarks on Logical Form."[25] But in fact there is no argument to be found in

---

[25]Wittgenstein, "Some Remarks on Logical Form," *Proceedings of the Aris-*

the *Remarks* moving from the analysis of color language to rejection of a primary or phenomenological language. In this particular regard the work in the *Remarks* is a radicalization of the position presented in "Some Remarks on Logical Form," where Wittgenstein appeals to the example of color propositions to show that there are certain propositions which are not themselves logically independent and yet cannot individually be analyzed into *independent* elementary propositions. Yet the very fact that, both in "Some Remarks on Logical Form" and in *Remarks*, Wittgenstein draws the conclusion that the *nature*, but not the existence, of elementary propositions must be rethought. This shows that at least in this context Wittgenstein does not use the points he makes about the logical dependence of color propositions to argue against a primary language.

Insofar as we are able to detect a heuristic motivation for rejecting such an appeal in Wittgenstein's consideration of color in the *Remarks*, it must lie in the recognition which Wittgenstein attempts to convey to the reader that visual impressions are much more complicated than they are usually taken to be. Thus, toward the end of the typescript Wittgenstein will remark that

---

totelian Society, Supplementary Volume 9 (1929), 162-71. There are some subtle problems of historical interpretation here that I am unable to resolve fully. But whatever the language of phenomenology (which in TLP is grounded in atomic propositions) is doing for Wittgenstein, it *isn't* about putting our language in order, since it's already in order in the first place. It seems that Wittgenstein *first* gives up the grounding of phenomenological language in atomic propositions, retaining the notion of phenomenological language. *Then* he gives up the language of phenomenology, seeing both that it isn't possible *and* that it isn't necessary. But the dilemma is: is it that it isn't necessary *because* our language is already in order? Wittgenstein had *already* recognized that our language is in order, so unless this was a "block" on his part, it seems this can't be the reason. I suggest below that Wittgenstein's new recognition involves seeing the lack of bearing atomic propositions have on our analysis of language. This does not resolve all the issues of interpretation, but it is a first step.

Nowadays the danger that lies in trying to see things as simpler than they really are is often greatly exaggerated. But this danger does actually exist to the highest degree in the phenomenological investigation of sense impressions. These are always taken to be *much* simpler than they are.[26] (PR, 281)

But at the very beginning of the typescript Wittgenstein simply *begins* with the assertion that "[a] proposition is completely logically analysed if its grammar is made completely clear: no matter what idiom it may be written or expressed in," and immediately follows this by remarking,

I do not now have phenomenological language, or 'primary language' as I used to call it, in mind as my goal. I no longer hold it to be necessary. All that is possible and necessary is to separate what is essential from what is inessential in *our* language. (PR, 51)

Thus the external division between a primary language and our language is replaced by a division within our language (or any other) between what is essential and what is inessential, and so the task of constructing a phenomenological language is replaced by the process of recognizing which parts of our language are wheels turning idly, which, as Wittgenstein somewhat cryptically says, *amounts* to the construction of a phenomenological language [*kommt auf die Konstruktion einer phänomenologischen Sprache hinaus*] (PR, 51). However we are to construe this last assertion,[27] it is impor-

---

[26]Ludwig Wittgenstein, *Philosophical Remarks*, trans. Raymond Hargreaves and Roger White, ed. Rush Rhees (Chicago: Chicago, 1975); hereafter cited as in text.

[27]It seems that the status of propositions accompanying philosophical activity changes from the *Tractatus* to the *Remarks*, and this bears on the status of phenomenological language, but the details ultimately remain unclear to me. The attempt to elucidate grammar propositionally seems to me to be the central concern.

tant to note that Wittgenstein continues to speak in the *Remarks* of phenomenological (or as he sometimes calls them, interchangeably, epistemological) concerns (PR, 88). In this way we are able to command an overview of grammar, and this allows us to separate what is essential in language (its grammatical structure) from what is not.

But why, then, are we to take this to be a problem internal to our language as we find it? Indeed, the problem is vexed in multiple regards, for in the *Tractatus*, where Wittgenstein is committed to a primary language of elementary propositions (without, on principle, telling which propositions these are or, indeed, giving us much sense of what they might look like, which can only be determined by the *application* of logic (5.557)), he nonetheless remarks as well at 5.5563 that "[i]n fact, all the propositions of our everyday language, just as they stand, are in perfect logical order." So, even in the context of the *Tractatus* the appeal to a language of elementary propositions cannot be understood as a requisite for *putting* language in logical order, since it is already in order in the first place. Indeed, in the context of the *Tractatus* Wittgenstein tells us that the attempt to say *a priori* what elementary propositions there are must lead to nonsense (TLP, 5.5571). The extent of writing on the *Tractatus* is enormous, but in my limited acquaintance with it I have not found that anyone has much of an illuminating nature to say about the role played by Wittgenstein's appeals to elementary propositions in the *Tractatus* (as distinct, that is, from talking about how Wittgenstein characterizes them). On the basis of the passages cited above, it is clear that, unless Wittgenstein has simply *forgotten* what he said about elementary propositions in the *Tractatus,* he could not *possibly* be concerned (either in "Some Remarks on Logical Form" or in *Remarks*) about whether color propositions would be candidates for elementary propositions, and recognition of this simple fact leaves behind much of what has been written about the role played by the consideration of color propositions in the turn in Wittgenstein's view. Further, it gravely distinguishes Wittgenstein's position from that of both Russell, on the one hand, and the logical positivists on the other. But, at the very least, we can say in what regard the situation changes from the *Tractatus* to the *Remarks*, and this is already something that has often been gravely mistaken: *not* in terms of putting versus finding our language in logical order, but rather in terms of what is possible and necessary in terms of so finding it: it is here that the position of the *Remarks* differs from that of the *Tractatus*. It seems that the most reasonable position to adopt is that in the *Tractatus* Wittgenstein viewed the availability of a fundamental language of description in terms of elementary propositions as a *consequence* of the fact that our language is in logical order: this is the sort of thing that a language which is in logical order should be susceptible of. Wittgenstein first step is then to question the grounding of such a fundamental language of description in atomic facts. Not only is an appeal to elementary propositions no longer a part of Wittgenstein's picture of logical or grammatical space: it is, indeed, neither necessary nor even possible, and so phenomenology, which consists in *seeing* the essential logical or grammatical features of language, must proceed without it– given that it isn't even possible, thank goodness that it isn't necessary! Wittgenstein's next step, which he reaches in the *Remarks*, is to doubt the existence of a fundamental language of description itself, asserting instead that a philosophical investigation "amounts" to such a description. Here, again, we can say that given that such a fundamental language of description is impossible, thank goodness it isn't necessary. Why, given that in each case it isn't possible, it isn't necessary has everything to do, I think, with Wittgenstein's commitment to the definiteness of logical form. I will have more to say about this below.

In fact, Wittgenstein does give us something like an argument for his new understanding of this state of affairs, thus supplying us with evidence of his lack of philosophical amnesia on this front.

Suppose there is an ideal, i.e. primary or phenomenological, language. What would such a language express? If it expresses what is expressed in our ordinary language, then why should we investigate this ideal language rather than our ordinary one? On the other hand, if it expresses something other than what our ordinary language expresses, how could I determine what this would be? As Wittgenstein puts it, "logical analysis is the analysis of something we have, not of something we don't have. Therefore it is the analysis of propositions *as they stand*" (PR, 52). As I have presented the position of the *Tractatus* above, this would be the case from the vantage of the *Tractatus* too once we see that, in fact, given the constraints laid upon elementary propositions there they can in fact have no *bearing* on our analysis of language. This is to say that, in one regard, there is a profound continuity between the projects of the *Tractatus* and the *Remarks*, insofar as ordinary language is recognized to be in perfect logical order *in both*. But the author of the *Remarks* is presumably castigating the author of the *Tractatus* for thinking that reference to elementary propositions and facts could be any part of the *representation* of logical space in accordance with this tenet.[28] Wittgenstein's progress should be viewed *not* as one of radical replacement of one philosophical project by another, but rather as the progressive *internalization* of the requirements imposed by the nature of the project itself.

A second argument which Wittgenstein presents early in the *Remarks* must equally be disambiguated. Because the text is difficult, I supply it here verbatim in my own translation:

> If I could describe the aim of the grammatical conventions by saying that I had to make them so because the colors have certain properties, then these conventions would be superfluous, for then I would be able to

---

[28]This perhaps already reflected in the *Tractatus* as well in terms of the problem associated with the status of objects: what could their status be other than logical? And yet logical constants do not refer.

say precisely what the conventions rule out [i.e. rule unsayable]. Conversely, if the conventions were necessary, so that certain combinations of words must be ruled out as nonsense, then I can precisely not provide a property of colors which makes the conventions necessary, for then it would be possible to think that the colors didn't have this property, and that could only be expressed contrarily to the convention. (compare the Hargreaves/White translation at PR, 53)

I understand this argument to say the following. If the grammatical conventions are a function of being constrained by the properties of colors themselves, then Wittgenstein tells us that paradoxically the conventions don't rule out precisely what the conventions themselves would dictate to be nonsense and, as such, the conventions have no effective power. On the other hand, if the conventions rule out certain candidate propositions as nonsense (and therefore not propositions, but simply word combinations), then this can't be a function of the way colors themselves are, since that is a contingent matter that we can imagine otherwise and so can't serve as a ground for ruling out certain grammatical constructions as *nonsensical*. This is all predicated, as I see it, on the deep commitment that sensicality is *not* a contingent matter.

In the *Tractatus* Wittgenstein had insisted that the atomic facts, and consequently the truth values of the elementary propositions, were a matter of empirical contingency: "each item can be the case or not the case while everything else remains the same" (TLP, 1.21). But if this is indeed so, then by the above argument, the truth values of the elementary propositions can contribute *nothing* (logically) to our understanding of the grammatical conventions of ordinary language. In particular, our presentations (*Darstellungen*) of grammatical space must come from elsewhere. Where, then, do these presentations come from, or is it even a mistake to speak this way?

A third passage with a similar argumentative structure to the two cited above may begin to help us with this question. Here, in the context of criticizing Russell's theory, Wittgenstein considers the relation of language to action and asks,

> If when a language is first learnt, speech, as it were, is connected up to action– i.e. the levers to the machine– then the question arises, can these connections possibly break down? If they can't, then I have to accept any action as the right one; on the other hand if they can, what criterion have I for their having broken down? For what means have I for *comparing* the original arrangement with the subsequent action? (PR, 64)

Wittgenstein then continues by asserting that "[i]t is such *comparison* which is left out in Russell's theory" (PR, 65). On Wittgenstein's new position, on the other hand, propositions are connected, as is exemplified in the case of color propositions, by certain propositions ruling out certain other propositions. What is crucial here, according to Wittgenstein is that, as he puts it, "I expect that the rod will be 2 m [meters] high *in the same sense* in which it is now 1 m 99 cm high" (PR, 65). That is, such propositions inhabit a common grammatical space which binds them together: asserting that a rod is 2 m high *means*, in part, that it is *not* 1 m 99 high. And the force of the argument given above is the same as in the two arguments cited previously: it is to show that grammatical space, and hence *meaning*, cannot be specified in terms of the relation of an ideal language to our own, of the properties of objects to the propositions which speak about them, or of speech to action. In each case the problem is the same, and leads to the same result: we look outside our language for an account of meaning, and in doing so we appeal to something which is neither a possible source for such accounting, nor is it needed.

It might be objected that, although in some rather weak and gutless sense, if such appeals are not possible then they cannot possibly be necessary, nevertheless this shows very little about how we *are* to provide an account of *Sinn*, sense or meaning, without such appeal. Indeed, I think we should not expect the argument style supplied in the above passages to provide much illumination in this regard: it is the major task of the *Remarks* to supply such an account and, hence, such an understanding, and these preliminary arguments are far from accomplishing that major task. I cannot hope, nor indeed is it in any sense my aspiration, to provide a full consideration of this project. Instead, my concern is merely to provide some suggestion for how the two large bodies of work upon which Wittgenstein focuses here, namely, the status of color propositions and the space of visual perception on the one hand, and the grammar of the infinite in a mathematical context, form a coherent whole for investigation. Here, my thesis can be stated, in first approximation, rather simply: it is that the set-theoretic appeal to a continuum of real-numbers is analogous to the appeal to an elementary or phenomenological language in the context of an account of language, and so should be rejected on similar grounds as both incoherent, i.e., impossible, and unnecessary. This is Wittgenstein's resolution of the Cantorian (set-theoretic) conception of the continuum. This interpretation has, I believe, several merits, but the first of these, if not the foremost, is that it offers an explanation for why Wittgenstein would focus so intensely on one particular problem of mathematical grammar in a context where his desire was to provide an account of the status of ordinary language from within language itself. Why, that is, is it the case that as always in the sphere of the infinite it is grammar which is playing tricks on us (PR, 208)? In particular, once we untangle this grammar of the mathematical infinite we are meant to see that a set-theoretic accounting for mathematics is just as needless as a Russellian theory of definite descriptions. Indeed I think this will be the case for Wittgenstein for just the same reasons, but for now I will leave this further contention aside.

There is a further issue which is puzzling: Wittgenstein gets rid of the notion of a set-theoretic continuum, but he seems dependent on talking about grammatical *spaces*. How are we to take this? One interesting suggestion is that *both* our notion of mathematical space and our notion of logical space are coming as a result of our perception of visual space and yet that in both cases we must equally not *derive* these spaces from visual perception. In any case, in both the logical and the mathematical sense Wittgenstein is dispensing with notions of space which are resolved down into independent atoms. What to make of this? My own view is that Wittgenstein's talk of grammatical spaces is *deeply metaphorical.*[29]

When Wittgenstein discusses grammar he often speaks in terms of the register of providing the requisite degrees of freedom.[30] The recourse seems to be to the analogy with a mathematical variable standing for a degree of freedom within a functional expression, and the point is that grammatical space is *not* given aggregationally. Color propositions serve as an example: saying that something is red all over means (in part) that it is not green all over. We may conjecture that part of Wittgenstein focus on color propositions is a function of his concern for how grammatical spaces "cohere." If this is right, then metaphorically at least this preoccupation is deeply connected to his castigation of the Cantor's aggregative conception of the continuum, and the elimination of elementary propositions from Wittgenstein's enterprise can be seen as a first step in eliminating a falsely aggregative, atomic conception of language.

---

[29]I discuss the consequences of such fundamental metaphorical recourse in *The Pace of Modernity: Reading With Blumenberg*, 21ff, and in *Diagnosing Contemporary Philosophy with the Matrix Movies*. On grammatical space as metaphorical, compare Robert Alva Noë, "Wittgenstein, Phenomenology and What It Makes Sense to Say," *Philosophy and Phenomenological Research* **54** 1 (1994), 1-42, 36. Noë compares this metaphorical sense to what Wittgenstein will come to call "being captured by a picture."

[30]Consider, in particular, the passages in which he talks about yardsticks at PR, 64, 65, 67, 68.

6.3.2 The (Cantor-Frege-Russell) Continuum According to Wittgenstein: A Commentary on Philosophical Remarks Section XV

In this subsection, I offer a detailed commentary on Section XV of Wittgenstein's *Philosophical Remarks*. In roughly ten pages Wittgenstein discusses the Cantorian continuum, Russell's use of the ancestral relation and, ultimately of most interest for my purposes, arguments for logical determinacy. What is of particular interest to me is that the arguments regarding logical determinacy (what this means I will explain in due course) come up so explicitly in precisely this context. My claim will be that Wittgenstein is investigating the grammar of the mathematical infinite at such length precisely because it proves to be such a difficult test of his commitment to logical determinacy. In order to maintain this commitment, Wittgenstein will be forced to adopt a perspective which will lead, ultimately, to the rejection of an independent phenomenological language as well. If my claim is right, it is the underlying commitment to logical determinacy which is driving the movement in Wittgenstein's thought in these other regards.

6.3.3 §170

Wittgenstein begins this section with a discussion of how "[t]he theory of aggregates attempts to grasp the infinite at a more general level than a theory of rules" (PR, 206). It will be the purpose of the remarks following this assertion to investigate just why there is no more such general level to be grasped, and so the picture with which Wittgenstein will leave us is of set theory as a black box into which a variety of mathematical rules are thrown, or equally of a gauze with which these rules may be covered. (Literally, Wittgenstein speaks of Russell wrapping up concepts in such a way that their form disappears.) Wittgenstein goes on to draw a distinction implicitly between description and presentation, and then characterizes the set-theoretic view as saying that the actual infinite can only be described and not presented since it can't be grasped by means of arithmetical symbolism at all.

In order to get a clearer picture of what Wittgenstein means, it will help to investigate the distinct uses to which he puts the terms 'description' (*Beschreibung*) and 'presentation' (*Darstellung*). In the *Tractatus*, Wittgenstein says that a picture *presents* a possible situation (*Sachlage*) in logical space (TLP, 2.202) and that a proposition *describes* a state of affairs (*Sachverhalt*) (TLP, 4.023). In the *Remarks* as well, '(re)presentation' will be used to refer to the domains of sense, possibility/necessity and internal relation, whereas 'description' will be used to refer to the domains of actuality and external relation. To say, then, that set theory says that the actual infinite can only be described and not presented means that the infinite is being taken extensionally, as an object. Wittgenstein's attitude toward the mathematical infinite, of course, will be quite otherwise.

Wittgenstein goes on to characterize the set-theoretic commitment to the actual infinite as analogous to buying a pig in a poke. We can't see it directly, but the description we give of it is "something like the way in which you carry a number of things that you can't hold in your hands by packing them in a box" (PR, 206).[31]

In the second paragraph Wittgenstein links this attitude to the

---

[31]compare Charles Sanders Peirce, "How to Make Our Ideas Clear," in *The Essential Peirce: Selected Philosophical Writings*, vol. 1 (1867-1893), ed. Nathan Houser and Christian Kloesel (Bloomington: Indiana, 1992), 124-41, esp. 132. Also compare Marcel Duchamp's *With hidden noise* in *Marcel Duchamp*, ed. Anne D'Harnoncourt and Kynaston McShine (Greenwich: MOMA/New York Graphic Society, 1973), 280, illustrated facing 288, and discussed in Carol P. James, "Duchamp's Silent Noise / Music for the Deaf," in *Marcel Duchamp: Artist of the Century*, ed. Rudolf E. Kuenzli and Francis M. Naumann (Cambridge: MIT, 1989), 106-26, 116.

one that would sanction the idea that logical forms can be described linguistically (rather than being presented). In this context Wittgenstein makes the problem (as he takes it) with such packaging clear: it is that the packaged, here concepts, rests on definitions, here the packaging; but if we unpack the packaging we find in the process that the concepts are themselves part of this packaging structure.[32] Wittgenstein does not focus on the circularity of this procedure, however, but rather on the way it obscures the concepts: we package them inside something which itself presupposes them, but in which, even more pressingly, the structure of these concepts is rendered amorphous. According to Wittgenstein, the point of this method is to make everything amorphous and treat it accordingly. By analogy, Wittgenstein's point about the Cantorian continuum would seem to be that it is mathematical packaging which presupposes the concept of mathematical rule but that renders the operative nature of this latter concept obscure. It should be remarked, however, that the way I have drawn out this analogy is crude, and what Wittgenstein goes on to say about both the continuum and logical form will ultimately require refinement.

Wittgenstein closes §170 with a short paragraph about generality and a second short paragraph in which his point is applied to a mathematical example. In the first paragraph Wittgenstein asserts what I will refer to as Wittgenstein's logical principle: in logic, the general case must always include the particular as a possibility. As an example, he gives the case of calculating a limit with two quantities δ and ν, which must include the number system of the particular computation. (This example suggests that this logical principle may be related to continuity principles, which have generally had a controversial status in mathematics. We have seen continuity priciniples enter into Leibniz's work at key points.) Put

---

[32]The circularity Wittgenstein identifies bears comparison with Isles' diagnosis of the circularity inherent in the typical presentation of the natural numbers, discussed above.

another way, a particular computation always occurs in the general context of a number system; as we will see, this example has considerable bearing on what Wittgenstein will go on to consider.

6.3.4 §171

Indeed, Wittgenstein begins the next section with the need to prove a function continuous within the context of a numerical scale–a number system. The claim here, as Wittgenstein goes on to specify immediately, is that any context in which I make a given δ depend on a given ν refers *ipso facto* to a general arithmetical criterion specifying how δ depends on ν so that $\varphi(\delta) <\nu$.[33] Wittgenstein's insight leads to a deep, but subtle, point: if we require the possibility of a numerical scale in the case of *particular* numerical calculations, we cannot eliminate it when we treat the *general* case. This is a direct consequence, indeed a simple application, of the logical principle announced above, but it means that any account of the continuum will require that the context of numerical scale *not* be eliminated. Insofar as the Cantorian continuum is viewed as an abstraction independent from such scaling, the set-theoretic approach does just that, and so it must fall to Wittgenstein's logical principle. This gives us all the more reason to evaluate Wittgenstein's logical principle carefully, but I will defer that task for the time being, since we will have a much richer context for this evaluation a little further on.

The direction in which Wittgenstein is driving at the beginning of this section is perhaps made most forceful in his remark that "[a] number system is not something inferior–like a Russian abacus–that is only of interest to elementary schools, while the higher, general discussion can afford to disregard it" (PR, 207). But it is just this tendency to think of the number system, and in particular

---

[33] I am following Wittgenstein's presentation here, which collapses some mathematical details.

decimal representation, as something inferior which leads to the tendency to stress a picture of the Cantorian continuum as general in a way which leaves behind just that concrete structure of the number system by way of which its construction is approached. Here we see an instance of the conviction Wittgenstein already expresses in the *Tractatus*, and which I have taken as an epigraph to this chapter: "(Our problems are not abstract, but perhaps the most concrete that there are)."[34] (TLP, 5.5563)

The next paragraph apparently signals an abrupt turn in Wittgenstein's train of thought: he asserts that the Cretan liar paradox, in which I assert 'I am lying' could also be set up by having someone write 'This proposition is false'. But the abruptness of this turn is only apparent, for Wittgenstein goes on to draw the general moral that "[t]he basic mistake consists, as in the previous philosophy of logic, in assuming that a word can make a sort of allusion to its object (point at it from a distance) without necessarily going proxy for it" (PR, 207-8). I take it that Wittgenstein's argument here is that both paradoxical sentences are seen to be resolvable in the same way once we see that the basic problem lies in failing to distinguish between the descriptive and the presentative function of words. A paradox is engendered in both cases if we take the subject term ('I'/ 'This proposition') to be descriptive without (necessarily, in these cases) *also* being presentative. The resolution of the paradox lies in the careful distinguishing of these dimensions. So the traditional problems surrounding the liar paradox illuminate the trouble with the Cantorian continuum (and, presumably vice versa, but more on that below). It is no surprise, then, that the next two short paragraphs, in their entirety, run:

> So the question would really be: Can the continuum be described? As Cantor and other tried to do.

---

[34]Mel Bochner has followed out these concerns as a visual artist; see Richard S. Field, *Mel Bochner: Thought Made Visible 1966-1973* (New Haven: Yale University Art Gallery, 1995).

> A form cannot be described: it can only be presented.
> (PR, 208)

We may summarize the position, to this point, as follows: set theory, and like it Russell's theory of definite descriptions in *Principia Mathematica*, make the basic mistake of trying to describe something that can only be presented. As such, these theories do not in any sense elucidate or ground what they would claim to package ("found"), but instead they have quite the opposite consequence: they obscure the form of that which they claim to describe.

In the last paragraph of this section Wittgenstein remarks that Dedekind's definition of an infinite set is another example of an attempt to describe the infinite, without *presenting* it. Dedekind's construction of the irrational numbers will be a subject of focus later on.

6.3.5 §172

In this section Wittgenstein gives an example of what it means to say that the continuum is *not* composed of points, and so of what it means to reject the Cantorian approach to the continuum. It becomes clear that this amounts, effectively, to the abandonment of an extensional treatment of the (actual) infinite, and so Wittgenstein shows us (by example) an alternative to such an extensional approach to the infinite. In this section, the refutation of the extensional approach, supplied in §170-171, is taken for granted, and so what we have is not an argument for Wittgenstein's alternative approach but rather an illustration of it. This illustration proceeds in terms of the example of the meaning of 'the highest point of a curve': it does not, Wittgenstein tells us, mean 'the highest point among all the points of the curve' since we aren't given all the points on the curve extensionally (or, as Wittgenstein puts it, we don't see these). Rather, it must mean a particular point on the curve such that given any (other) point on the curve it is higher than that (presumably Wittgenstein here intends a strict maximum). The point is that other points are *characterized* rather than the totality of them being given extensionally, and like the specification of the value δ in terms of the value ν in the example given above, here Wittgenstein indicates a *logical relation* between points. Similarly, Wittgenstein says, the point of intersection of two lines isn't the intersection of two classes of points but rather the meeting of two laws, i.e. the condition under which the laws are jointly satisfied.

Wittgenstein goes on to make it explicit that we must distinguish grammatically between two different things: the highest point being higher than every other point, and the highest point being the maximum value in an extensional collection of values. Again it is grammar which, as always in the sphere of the infinite, is playing tricks on us, and it is by providing an intensional account of the mathematical infinite that these tricks can be resolved. Wittgenstein indicates that we must also distinguish between the grammar of the infinite and of the finite: the expressions 'divisible into two parts' and 'divisible without limit' have completely different forms; failing to distinguish between them is tantamount to treating 'infinite' as if it were a number word and, in so doing, failing to distinguish between these forms. Our motivation for doing this, as Wittgenstein points out, is that in everyday speech, both are given as answers to the question 'How many?', but this grammatical conflation is an example of just the sort of tricks which grammar plays on us.

On Wittgenstein's intensional perspective "[t]he curve exists, independently of its individual points" (PR, 209). That is, the curve isn't composed of points at all. The points, rather, such as the maximum, are constructed in the sense that we derive them from a law and not by examining individual points. Equally, we don't characterize the intersection of two lines as one among *all* the points on a line: "no, we only talk about *one* point." Hence, here too: "[t]he straight line isn't *composed* of points" (PR, 209).

6.3.6 §173

So far, Section XV is certainly challenging, but nonetheless fairly straightforward going; the same cannot be said of the material in §173. Nonetheless, it is indeed critical for Wittgenstein's train of thought and cannot be avoided. I simply note that more connections must be supplied and so the reconstruction must be more tentative. Also, in order to aid in reconstruction I will return to the manuscript context at WA, I, 128ff.

This number in PR begins with a reference to the ancestral relation as developed by Russell,[35] and passes immediately to a remark about the untenability of Ramsey's account of the infinite.[36] The manuscript evidence makes clear that this train of thought occurs in the context of a consideration of the nature of mathematical generality and, in particular, the way in which one instance may be said to contradict a general proposition. (There is some discussion of generality in Section XIV of *Remarks*). The basic problem here

---

[35]The notation given in the printed version of *Philosophical Remarks*, 'R*', is inaccurate, and the *Wiener Ausgabe* confirms that Wittgenstein correctly wrote 'R_*' instead (WA, I, 128).

[36]Ramsey's program to eliminate Russell's appeal to the Axiom of Reducibility is given in Chapter One of *The Foundations of Mathematics and Other Logical Essays*, ed. R. B. Braithwaite (London: Routledge, 1931). It relies heavily on an appeal to the Axiom of Infinity, as is stressed in the discussion of Bernays in "The Philosophy of Mathematics and Hilbert's Proof Theory," translated in Paolo Mancosu, *From Brouwer to Hilbert: The Debate on the Foundations of Mathematics in the 1920's*, 234-265; here see 256-7. The material about Cantor and the theory of aggregates occurs in the *Wiener Ausgabe* circa I, 122. This edition confirms that it is to R_* that Wittgenstein refers and not to R*, as printed in *Remarks*; see WA, I, 122. The material concerning the ancestral relation in Russell and Whitehead's *Principia Mathematica* is given in volume I Section E at *90. An introductory discussion is given in the Sumary of Section E at 543-8. Here Russell acknowledges that "[t]he present section is based on the work of Frege, who first defined the ancestral relation" (548n). References are given to the *Begriffschrift* Part III 55-87 and the *Grundgesetze der Arithmetik* I §§45-6.

is adumbrated in passages that appear in *Remarks* in Section XII:

> Yet it still looks now as if the quantifiers make no sense for numbers. I mean: you can't say '$(n) \cdot \phi n$', precisely because 'all natural numbers' isn't a bounded concept. But then neither should one say a general proposition follows from a proposition about the nature of number.
>
> But in that case it seems to me that we can't use generality – all, etc. – in mathematics at all. There's no such thing as 'all numbers', simply because there are infinitely many. And because it isn't a question here of the amorphous 'all', such as occurs in 'All the apples are ripe', where the set is given by an external description: it's a question of a collection of structures, which must be given precisely as such. (PR, 148)

I understand this passage as follows. First, quantification over numbers cannot make sense, at least in the way we use 'all' when we quantify over a collection of apples, because we cannot think of 'all natural numbers' on analogy with 'all apples' (on this table, say). A proposition about the nature of number cannot be thought of as general on analogy with the generality of 'All apples are ripe', and so neither can anything that follows from it. When we are making propositions about the nature of numbers we are referring to structures or collections of structures, and not collections of numbers as objects given extensionally. Second, we must distinguish between saying, e.g., 'The numbers from 1 to 10 are 10 in quantity' and 'There are infinitely many numbers'; the latter proposition does *not* assign a quantity to the collection of all natural numbers. (In this regard, Leibniz's account of the mathematical infinite as indefinite may be seen as a precursor.) Strictly speaking, there is no collection of all natural numbers, but only, we might say, a structure of natural numbers. Describing this structure is not an abstract problem of generality, but is perhaps among the most concrete there are.[37]

Russell's treatment of Frege's ancestral relation (and presumably Wittgenstein would level the same criticism of Frege) is amorphous because it is aimed at giving an account of number in general. I take it that when Wittgenstein asks what a correct, as opposed to an amorphous explanation of R* [sic] would be like he is asking how or whether the ancestral relation can be squared with the recognition about the absence of generality expressed in §126 in order to give an account of finite number and, hence, quantity. Wittgenstein begins by recognizing that he will need the quantifier '(n)' which expresses the existence of n things satisfying a given predicate, but seems to think that in this context such a (bounded) quantifier is admissible. However this may be, and we will see in a moment that Wittgenstein recognizes there are problems, the quantifier '$(\exists x)$', however it is to be understood, cannot be construed in a way which would presuppose the totality of numbers, since this would reduce it to nonsense.

Ramsey's enlistment of the ancestral relation in his account of the infinite is equally nonsense, for it would presuppose that we were given the actual infinite and not merely the unlimited possibility of going on (PR, 209-210); this, it seems, is not a consequence of using the ancestral relation itself, but rather of the use to which Ramsey puts it. And yet in a note the editor Rush Rhees points out that in the manuscript source Wittgenstein then wrote that he questions the logical coherence of the ancestral relation itself because the variable n needs to be bounded both from above *and* from below.[38] The general problem, however, is clear: if we are to

---

[37]See also WA, I, 117 entry 1: "In keiner religiösen Confession ist soviel durch den Mißbrauch metaphorischer Ausdrücke gesündigt worden wie in der Mathematik" [In no religious confession has there been so much sin committed through the misuse of metaphorical expressions as in mathematics]. This makes the question of the literal versus metaphorical status of grammatical space, as discussed above, all the more pressing.

[38]see WA, I, 128. The passage in which Wittgenstein makes this remark is

speak of quantification in a numerical context at all, we must not do so in such a way that it would require reference to the totality of all numbers. In a second manuscript passage which Rhees includes in a note, the point is made that if the quantifier $(\exists nx)$ (there are n xs) is admitted then it will be possible to form the quantifier (n) (which quantifies over predicates) and Wittgenstein claims that this presupposes that there are infinitely many objects. This passage is perhaps omitted from the typescript itself because Wittgenstein is uncertain whether the ancestral relation, and hence the quantifier (n), is logically coherent or not; in the text itself Wittgenstein restricts himself to posing the question whether it is acceptable to express a one-one numerical correspondence between two predicates ϕ and ψ by '$(\exists n){:}(\exists nx){\cdot}\phi x \cdot (\exists nx){\cdot}\psi x$'. In the typescript, then, the status of the quantifiers '$\exists nx$' and (n), and hence their ultimate relation, appears to be left open.

Wittgenstein turns next to the status of Brouwer's pendulum number, about which we are unable to say whether it is equal or not equal to the number 0.[39] Wittgenstein asserts that the fact that such numbers are incompatible with the law of excluded middle doesn't reveal a peculiarity of propositions about infinite aggregates, but rather is based on the fact that logic presupposes that it cannot be *a priori* – i.e. logically – impossible to tell whether a proposition is true or false. Wittgenstein must be committed to this position, typically associated with espousal of the law of excluded middle, since the sense of a proposition is in principle a function of its truth or falsity and so a failure of excluded middle would force the loss of sense of the candidate proposition. That is, Wittgenstein is committed to what I shall refer to as a principle of logical definiteness: that a proposition has a sense if (and

itself marked with the symbol '*'.

[39] For a good discussion of Brouwer's pendulum number and its role in constructing weak counterexamples, see Mark van Atten, *On Brouwer* (Toronto: Wadsworth, 2004), 26-9; I return to the relation of Wittgenstein and Brouwer below when I discuss Mathieu Marion's interpretation of Wittgenstein.

only if) it is in principle assignable a truth value (either true or false). Wittgenstein is committed to a very strong form of bivalence here, and in this sense he is as far as imaginable from the intuitionist position. *Anything not subject to the principle of logical determinacy– here specifically with respect to determinate truth values–Wittgenstein takes by that token to be nonsense.*

Wittgenstein objects to the division into a finitary and an infinitary domain internal to mathematics on just this ground of logical determinacy: since truth and falsity are logical statuses, something is either true or false *simpliciter* so that, as Wittgenstein goes on to say at the beginning of §174, "[t]here is no such thing as an hypothesis in logic" (PR, 211). Although such a figure as Brouwer (or, presumably, Weyl) holds the division of mathematics into finitary and infinitary regions "to be particularly subtle, and to combat prejudice" (PR, 211), according to Wittgenstein such a view "is quite out of place in mathematics, completely contrary to its essence" (PR, 210). Matters of essence are matters of logic or, as Wittgenstein will say with increasing frequency from the time of the *Philosophical Remarks* onward, grammar.

§173 closes with a return to the pernicious idioms of set theory and a brief discussion of Dedekind. In both cases the target is a symbolism which would make us believe we may coherently think of something which is, in fact, nonsense. Speaking, as one often does in a set-theoretic context, of a line as composed of points deflects attention from the fact that "[a] line is a law and isn't composed of anything at all" (PR, 211). This leads, in particular, to (false) surprise that there would be room between rational points for irrational points to be interposed. Rather, a construction like that for $\sqrt{2}$ merely shows that the point *yielded* by the construction is *not rational* (PR, 211). Analogously, in the context of Dedekind's construction of the real numbers, Dedekind "proceeds as if it were clear what was meant when one says: There are only three cases: either R has a last member and L a first, or, etc. In truth none of these cases can be conceived (or imagined)" (PR, 211). The point is made with full force at the beginning of §174:

> Set theory is wrong because it apparently presupposes a symbolism which doesn't exist instead of one that does exist (is alone possible). It builds on a fictitious symbolism, therefore on nonsense. (PR, 211)

6.3.7 §174

The above remark, which begins this number, is followed by the remark quoted above in which Wittgenstein asserts that "there is no such thing as an hypothesis in logic" (PR, 211). This latter phrase has made it difficult, in particular, to understand the status of a phenomenological investigation, which many commentators have understood as existing somewhere midway between an a priori logical investigation and an a posteriori scientific investigation. In particular, if it were indeed the case that the nature of elementary propositions were a matter for empirical investigation, then there would be no logical place for the investigation of a phenomenological or primary language. These issues, then, link the discussion in Section XV directly to issues regarding the status of a phenomenological language. But in the specific context of the discussion in Section XV, the emphasis is on problems surrounding the status of mathematical propositions and, in particular, the difficulties regarding what logical role quantification could play (if any) in this language. We have the dilemma, adumbrated above, that on the one hand mathematical propositions appear to require quantification over collections of numbers (if not, indeed, all numbers, which Wittgenstein finds incoherent) but, on the other hand, it is not clear either, as is indicated in the manuscript sources, that even quantification over bounded collections, understood in terms of Frege and Russell's ancestral relation, can accomodate such quantification coherently.

Wittgenstein makes a remark, analogous to those in the last number, that distinguishes between saying 'The set of all transcendental numbers is greater than that of algebraic numbers', which is nonsense, and saying that the set of transcendental numbers is *not* denumerable, i.e. is of a different kind. He then goes on to make a rather cryptic remark that "[t]he distribution of primes would then for once provide us with something in logic that a god could know and we couldn't" (PR, 212). This is a case in which it is much easier to interpret Wittgenstein's intent by consulting the context provided by the original manuscript: there it is clear that Wittgenstein is describing the counterfactual situation which would obtain, and about which he explicitly indicates that he finds it unbelievable, were there to be a duality: law and the infinite series which follows from it, which is to say, were there to be a distinction in logic between description and reality (WA, I, 96 entry 9). In that (unbelievable) case we would have to recognize a knowledge which a god could have and which we cannot.[40] Once digested, these remarks provide a context for the discussion that will take up the rest of §174, of the highest importance for the understanding of the view Wittgenstein is presenting, surrounding problems associated with the idea of undecidability in mathematics and the (incoherent) sense of logical indeterminacy which it presupposes.

Wittgenstein begins this discussion by asserting that it cannot be asserted that there is a process of solution (of a mathematical equation). The equation depends for its sense on the possibility of checking whether it is true or false; if there were *no* process of solution then the equation would lack sense, and according to Wittgenstein's conception of the logical, where it is nonsensical to assert a proposition it is equally nonsensical to assert its negation.

---

[40]In the manuscript passage the passage included in PR ends with an exclamation point which has been changed to a final period in PR. There follows then in parentheses: "(Das ist es was ich nicht glauben kann)" (WA, I, 96 entry 10).

Wittgenstein concludes this passage by asserting that "[w]e can assert anything which can be checked in practice" (PR, 212). We cannot, however, *check* whether there is a process of solution for an equation, except of course by solving it, and that solubility is what *would* be in question according to the illicit proposition which Wittgenstein says cannot be asserted. I take it that Wittgenstein is distinguishing here between checking something, the possibility of which serves as a warrant for assertion, and incoherent attempts to *express* this warrant. We may put the point another way: what the illicit proposition attempts to express incoherently attempts to capture a general condition of the sense of an equation. It is precisely *not* that the claim that there is a process of solution for an equation is false; it attempts, rather, to say something which can only be shown by the sense of the equation. Consequently, Wittgenstein goes on to say that "[i]t's a question of the *possibility* of checking" (PR, 212, my emphasis).

Part of Wittgenstein's concern with such incoherent assertions is that they provide prima facie support for the idea of an undecidable equation, and so Wittgenstein turns to Brouwer's claim that for the equational proposition '$(x) \cdot f_1x = f_2x$' there is, in addition to the answers 'yes' and 'no', additionally the case of undecidability. Wittgenstein claims that this commits Brouwer to an extensional view of logic, and thinks that on such an extensional view the prospect of a question which is in principle undecidable would make sense– if, for example, mathematics were the natural science of infinite extensions (PR, 213). However, he is consistently at pains to dispute the coherence of just this picture, and the rest of the remarks in §174 combat this conception by arguing against the incoherent picture of undecidability which is its consequence.

For Wittgenstein, what allows us to *understand* an equation involves a connection between the two sides of the equation which is established by the function of the symbols on the two sides of the equation, and the nature of this connection must be such that it would be possible to decide whether the equation is or is not satisfied. Undecidability, on the other hand, "presupposes that there is, so to speak, a subterranean connection between the two sides; that the bridge *cannot* be made with symbols" (PR, 212). The idea of a symbolic connection that cannot be represented symbolically is a logical contradiction in terms:

> A connection between symbols which exists but cannot be represented by symbolic transformations is a thought that cannot be thought. If the connection is there, then it must be possible to see it. (PR, 212)

Wittgenstein follows this with a claim that the connection *exists* in the same way as a connection in visual space, and that neither is a causal connection. In each case the transition– in the mathematical case from one side of the equation to the other, in the visual case from one part of visual space to another– is produced without recourse to some kind of "dark speculation different in kind from what it connects" (PR, 213). The full implications of this analogy are not entirely clear to me, but two things are obviously important. The first is simply that Wittgenstein is connecting his mathematical analysis to the analysis of visual space, and in both cases is understanding the nature of the connection in such a way that it will be amenable to a logical, or, as he will also say, a grammatical analysis. The second is that in the mathematical case it is the idea of numbers in extension to which (attempted) recourse is made in the sort of dark speculation to which Wittgenstein refers.

I will return to this set of concerns in particular later in this chapter. To anticipate, I think that Wittgenstein fails to distinguish between the logical indeterminacy to which an extensionalist position leads and the commitment to logical determinacy which is required on his intensionalist intepretation of the mathematical infinite. Indeed, I will argue that the commitment to logical indeterminacy, while requisite on the extensionalist picture, as

194

Wittgenstein suggests, is also *available* as an alternative on the sort of intensional picture of logic which Wittgenstein espouses; it is this option which Wittgenstein leaves unconsidered.[41]

6.3.8 §175

This number largely revolves around questions about the incorporation of equations into propositions. The problem with the incorporation of equations into propositions is that if the equation does not have a solution, or even a solution appropriate to the context of the proposition, then the incorporation of the equation renders the proposition nonsensical. Wittgenstein first points out how the incorporation of equations into propositions in general generates a notation in which we cannot immediately tell sense from nonsense, and concludes further along that "I mustn't chance my luck and incorporate the equation in the proposition, I may only incorporate it if I know that it determines a cardinal number, for in that case it is simply a different notation for the cardinal number" (PR, 213). Otherwise the inclusion is random and leaves to chance whether there is any sense or not. Wittgenstein's insistence on viewing these simply as different notations for the same cardinal number is designed to preclude the problem which would be associated with viewing a phrase like 'the root of the equation $\phi x = 0$' as a Russellian description, for proceeding in this way different expressions for the same cardinal number would lead to the propositions in which they appear having different senses. Further, the correctness or incorrectness of equations is at issue in the same way when variables are involved in equations as when they are not: Wittgenstein gives as an example that $2 + n = 1$ is just as incorrect as $2 + 3 = 1$, while $2 + n \neq 1$ is perfectly correct.

---

[41]Michael Dummett's paper, "The Justification of Deduction," in *Truth and Other Enigmas* (Cambridge: Harvard, 1978), 290-318, provides a useful vantage.

6.3.9 §176

In this number Wittgenstein faces two related problems: first, the problem that a purely internal generality such as is expressed by an equation may be refuted by a single case, and, second, the problem that working over an incorrect equation to determine that it is incorrect involves an application of rules and seems to show us something about the general form of the (incorrect) equation. In the first case Wittgenstein refuses to acknowledge that there is a problem, remarking that the individual case refutes the generality from within, and not like the way in which a one-eyed man refutes the claim that all men have two eyes. In the second case, Wittgenstein appeals to a geometrical analogy to suggest that the manipulation of the equation may show us a connection between the forbidden equations and equations with normal solutions. Hence if $\sqrt{-1}$ is the result of such manipulation, "I at least know that $\sqrt{-1+1}$ would be a normal root" (PR, 215); Wittgenstein goes on to ask if this means that the concept of the imaginary numbers is already prefigured in the concept of the real numbers. This is interesting, in particular, because it suggests that Wittgenstein is willing to consider a connection between forbidden and permitted equations, so long as it is one which is made in terms of the extensions of their validity, as is equally the case in distinguishing the equations $x^2 = x \cdot x$ and $x^2 = 2x$ (PR, 215).[42]

So concludes Wittgenstein's Section XV. The concerns Wittgenstein expresses are part of the larger texture of the work, extending back from and beyond Section XV. But a reasonably definite picture of Wittgenstein's views of the Cantorian continuum emerges from a careful consideration of Section XV, and one which generalizes to Wittgenstein's views about the nature of logical definiteness. It is this connection between the specific investigation of the Cantorian picture of the continuum and Wittgenstein's underlying

---

[42]On numbers in elementary propositions, see WA, I, 63 entry 5.

logical commitments which, more than anything else, suggests why his preoccupation with the distinction between the finite and the infinite was one of the two driving forces in the transformation of his overall self-conception of his philosophical program. In terms of the larger program of this book, we may draw the following moral from the commentary I've supplied in this section: Wittgenstein's deep and unchanging commitment to logical definiteness led to his rejection of an extensional conception of the mathematical infinite and his cultivation of an account of the finite/infinite distinction as part of a larger network of grammatical structure. Concomitant with, but not, I will argue, reducible to this commitment to the rejection of extensionalism on logical grounds is Wittgenstein's commitment to the full determinacy of the logical structures he investigates. In Appendix D I compare how two other commentators on Wittgenstein's work have attempted to understand the transformation of Wittgenstein's thinking on these issues. This material most naturally fits at this point in the chapter, so those readers interested in it may wish to turn to Appendix D at this point.

6.4 Prospectus: Ethics and the Parafinite

6.4.1 Parafinite

In this final section, I will accelerate considerably and lay out a program for further development. Although it is based on what I have written above, it is by no means fully justified by it. As such, this section is not in any traditional sense a conclusion, but rather a prospectus for future investigation.

This prospectus is divided along the lines indicated by the heading of this section, and I will consider first Wittgenstein's views on the mathematical infinite. In this regard, I suggest that we may outline the position toward which Wittgenstein is moving in the following terms. Infinity is to be understood in terms of the appeal to a law. Laws are hypothetical. The language of phenomenology is the language of primitive (pure) assertion (contrary of hypothetical); thus the rejection of a phenomenological language is concomitantly the rejection of a language unaffected by law, by hypothetical form. With this rejection, we recognize that there are no pure assertions, and the distinction between hypothesis and assertion is relativized. This intensifies the need to find a way to describe how we find a rule followed in experience, and this must be reflected in the grammar of rule-following. What may be most controversial in the picture I've proposed is the role of the hypothetical, but I also think the picture can be largely reframed without this emphasis.

Take the rule which specifies the infinity of natural numbers in terms of succession. How do we know the character of this rule? John Mayberry has objected, in this context, that independently of a proof of the validity of an induction we have no grounds for characterizing the rule as definitely this or that. Wittgenstein seems to assume such a definiteness without proof.[43] Quoting Marion: "[i]f the role played by the expression 'and so on' is to show the possibility of the symbolism, it has, so to speak, a completely different, 'strict and exact' grammar" (WFFM, 182).[44] But why should we assume that this grammar is strict and exact? We find a similar claim made with respect to the transparency of the uniqueness of definition by recursion. In *Philosophical Grammar* (hence later)

---

[43]See, e.g., John Mayberry, *The Foundations of Mathematics in the Theory of Sets* (Cambridge: Cambridge University Press, 2000), 271n.7. I am far from being in agreement with Mayberry that Wittgenstein's account of number in the *Tractatus* is a farrago, but his central point still stands.

[44]Marion cites *Wittgenstein's Lectures, Cambridge 1930-1932* (Oxford: Blackwell, 1980), 89.

Wittgenstein replaces the rule of induction by the rule of inference:

$$F(0) = G(0),$$
$$F(Sx) = B(F(x)), G(Sx) = B(G(x)) \vdash F(x) = G(x).^{45}$$

This indicates that, at a minimum, and even on Marion's interpretation, what fundamentally distinguishes Wittgenstein from strict finitism is the commitment to the uniqueness of the natural number series. This may presumably be attributed to Wittgenstein's commitment to logical definiteness.

Marion gives as the features from strict finitism (SF) of interest to the interpretation of Wittgenstein, first, the critiques of the assumptions : (A1) The uniqueness up to isomorphism of the natural number series, and (A2) The principle of mathematical induction from n to Sn, and, finally, the introduction of the notion of 'feasible' number (WFFM, 214). But I think that Mayberry is right to say that the relevant critique of induction is already to be found espoused in Dedekind and, as such, is no part of an exclusive concomitant of strict finitism.[46] Mayberry would endorse the critiques associated with (A1) and (A2) but would not endorse the introduction of feasible numbers. Also, it is disastrously misleading to characterize the Wittgensteinian notion of infinity as law in terms of the Aristotelian notion of potential infinity, and so Marion's argument that Wittgenstein's interest in the surveyability of proof could not have originated in his recognition that 'human feasibility' must play a central role in philosophy and foundations of mathematics is vitiated, at least on this count.[47] This holds

---

[45]Ludwig Wittgenstein, *Philosophical Grammar*, trans. A. Kenny (Berkeley: California, 1974), 397, 414; see discussion in WFFM, 107.

[46]Mayberry, *Foundations*, 271-3. I have elaborated on Mayberry's discussion in my review of his book, published in *Notre Dame Journal of Formal Logic* **46** 1 (2005), 107-125.

[47]WFFM, 220. There is some manuscript evidence which may independently support Marion's point, but I leave aside this question of interpretation.

whether the conclusion is correct or not– I am more sympathetic to what Marion suggests about something ceasing to be a calculation (WFFM, 223). W. W. Tait has remarked, suggestively, that necessity rides on the back of contingency for Wittgenstein, but I think that these issues are integral to foundations of mathematics and are not comfortably compatible with the definiteness about rules that Wittgenstein insists on. Further elaboration and defense of this claim must await another occasion.

Here is the basic dilemma: it can perhaps best be couched in terms of Emile Borel's claim (which I take to be a potent one, despite the fact that it ignores crucial issues depending on the distinction between epistemological and logical concerns) that sufficiently large (finite) numbers have more in common with the infinite than with smaller finite numbers. For Wittgenstein the distinction between finite and infinite is categorical, and Wittgenstein has little to say about what distinguishes small from large numbers except that there is an override on rules which is dictated by our (changing) capacities for regularity of calculation. This makes it sound as if there is nothing which distinguishes simple calculations from complex calculations other than our cognitive capacities, and this hangs together well with Wittgenstein's resolute dissociation of the logical and the calculational. I have criticized this dissociation elsewhere, and what Michael Dummett has to say in "The Justification of Deduction" bears on this issue as well.[48] Wittgenstein's dissociation of logic and calculation drives away from the commitment to the parafinite as the logically large as you please insofar as it handles largness in terms of the cognitive override on the application of rules in complex conditions. There is some support for this position in Wittgenstein nonetheless, insofar as he insists that there is no sense in which the general can be independently "abstracted" from individual (structural) cases.

---

[48] "Miscalculation and Logical Validity," unpublished; also Michael Dummett, "The Justification of Deduction."

200

6.4.2 Ethics

The focus on the status of the indefinitely large may actually push forward some suggestions Wittgenstein makes, albeit tentatively, about how philosophical investigations into the foundations of mathematics bear on a certain crisis which he discerns in the *practice* of mathematics. The ethical attitude implied by Wittgenstein's remarks is severe, and not one I would endorse. However, it points the way to thinking about a set of issues which such ethical rigorism resolutely faces.[49]

Above I mentioned Borel's suggestion that sufficiently large numbers have more in common with the infinite than they do with "small" numbers. In line with this suggestion, consider the following passage from a paper by Kreisel, Mints and Simpson:

> For what it is worth our guiding idea is this. To master large finite configurations one simply must not (despite the stress that ideological advocates of *finitism* put on the possibility of such a step-by-step build-up (e.g. of the power set of a finite set), on the fact that the configuration *can* be so built up (in principle)) use their build-up!– and if one doesn't, the results will automatically apply to suitable infinite configurations.[50]

This suggests the slogan that infinitary mathematics should be viewed as finite mathematics *minus* an appeal to the hereditary construction of numbers from below. As such, infinite mathematics would be an approximation to finitary mathematics and not vice versa.

---

[49]For an extended consideration of the relation of Wittgenstein's logical and ethical views stressing Wittgenstein's religious orientation, see Philip R. Shields, *Logic and Sin in the Writings of Ludwig Wittgenstein* (Chicago: Chicago, 1993).

[50]G. Kreisel, G.E. Mints and S.G. Simpson, "The Use of Abstract Language in Elementary Metamathematics: Some Pedagogic Examples," *Logic Colloquium*, Springer Lecture Notes in Mathematics 453, 38-131, here, 70-1.

Ultimately I believe such a reversal of polarity is required in order to address issues of mathematical praxiology, i.e. questions about what mathematician's should do, and in a way which connects the concerns of the previous subsection directly with this one.[51] I offer two perspectives for consideration. First, a pair of passages from the work of Petr Vopěnka:

> [1] Set theory opened the way to the study of an immense number of various structures and to an unprecedented growth of knowledge about them. This caused a scattering of mathematics. Moreover, most results of this kind derive their sense only from the existence of the respective structure in Cantor set theory. Mathematics based on Cantor set theory changed to mathematics of Cantor set theory.
>
> Cantor set theory is responsible for this detrimental growth of mathematics; on the other hand, it imposed limits for mathematics that cannot be surpassed easily. All structures studied by mathematics are a priori completed and rigid, and the mathematician's role is merely that of an observer describing them. This is why mathematicians are so helpless in grasping essentially inexact things such as realizability, the relation of continuous and discrete, and so on.
>
> [2] Restriction of mathematical problems and concentration on problems posed by the alternative [i.e. Vopěnka's] theory is drastic. The spirit of the alternative set theory seems to regulate mathematical problems in medias res. To solve such problems is by no means easy. If our way or a similar way were shown

---

[51]This should not be confused with the distinction between praxiological and ontological *foundations* for mathematics, although it is obviously related to it in important ways.

to be right, then this fact would probably lead to a considerable restriction of mathematical production.[52]

Perhaps the most basic inexact notion in mathematics is the notion of numerical largeness, and Vopěnka's remarks apply *a fortiori* in this case. Although in a negative light, Badiou recognizes it in the passage I've taken as the third framing passage in the Preface to this book. But the status of numerical largeness must also be distinguished from the ethical rigorism which would promote the restriction of mathematical production, and which Wittgenstein shares. These two issues are related historically and conceptually, but their relations need to be articulated and the issues disentangled.

Next consider Wittgenstein, commenting on Hilbert's dictum that "No one is going to turn us out of the paradise which Cantor has created":

> I would say, I wouldn't dream of trying to drive anyone out of this paradise. I would try to do something quite different: I would try to show you that it is not a paradise– so that you'll leave of your own accord. I would say, You're welcome to this: just look about you.[53]

And later in the same lecture series,

> I am *not* saying transfinite propositions are *false*, but that the wrong pictures go with them. And when you see this the result may be that you lose your interest. It may have enormous consequences but not mathematical

---

[52]Petr Vopěnka, *Mathematics in the Alternative Set Theory* (Leipzig: Teubner Verlagsgesellschaft, 1979), 9 and 14 respectively.

[53]Ludwig Wittgenstein, *Wittgenstein's Lectures on the Foundations of Mathematics Cambridge, 1939*, 103.

consequences, not the consequences which the finitists expect.[54]

In these passages Wittgenstein's conception of mathematical consequences is precisely a conception of what would bear on the truth and falsity of mathematical propositions, so that questions about where mathematicians should turn their attention are not in this sense at all mathematical, but *rather* ethical ones. I view this dissociation of mathematical truth from the ethics of mathematical practice as an unattractive feature of Wittgenstein's view.

Perhaps most potently, Wittgenstein makes the following remarks, which appear in cipher script at around the time of the bulk of the material discusses from Section XV of PR:

> Ich glaube die Mathematik hat im vorigen Jahrhundert eine ganz besonders instinklose Zeit gehabt an der sie noch lange leiden wird. Ich glaube diese Instinklosigkeit hängt mit dem Neidergang der Künste zusammen, sie entspringt der selben Ursache.

> I believe mathematics went through a period of loss of instinct during the last century from which it will continue to suffer for a long time. I believe this loss of instinct is connected to the decline of the arts, it arises from the same root. (WA, I, 157)

I will have something to say below about whether this development, which I agree with Wittgenstein is quite real, should be regarded as a "crisis" or not. My suggestion in any case is that we may begin to redress the condition Wittgenstein points to, in its philosophical part, by considering the status of large numbers and the consequences of the way in which we consider them for the conduct of mathematics (and philosophy) at large.

---

[54]idem, 141.

This suggestion faces several immediate, and drastic, challenges. The first, and most obvious, is that it is *impractical*, if not indeed ridiculous, to think that in the current culture of mathematics or any differing from it in an evolutionary rather than revolutionary fashion, such severe restrictions of mathematical production could possibly be imposed. In a descriptive sense, Wittgenstein's prognosis that the "loss of instinct" in the mathematical community will continue for a long time has certainly been sustained. We may not agree that what Wittgenstein identifies was caused by a loss of instinct, but the massive proliferation of mathematical production is a cultural fact beyond dispute, and it shows no signs of abating. Furthermore, although Wittgenstein would presumably view his program of ethical reform as one cultivating a return to "instinct," I think many would view such a return with some horror, given the atrocities that were committed in the twentieth century in such terms.

The proliferation of mathematics is indeed bewildering, although its sprawl may also be overstated, and we should not immediately identify it with a sort of cancerous metastasis. Here is a vital counter to Wittgenstein's suggestion, given by the contemporary mathematician Alain Connes:

> I have a somewhat different point of view.[55] The increase in the number of mathematican not only had the effect of multiplying the number of interesting results, but it also facilitated the task of digesting these results. It seems to me that a huge number of articles are written just to simplify things and facilitate the work of those who wish to master a large area of mathematics. Whereas it would be impossible to read the countless articles in all mathematical journals, this work of selection of information is being accomplished better and

---

[55]Connes is replying to remarks by M. P. Schützenberger.

better, despite the background noise that has invaded the mathematical circus. There are perhaps 15,000 to 20,000 mathematicians. They are distributed among a certain number of villages. In any randomly selected village, the number of important advances made in four years separating two International Congresses of Mathematicians (ICM) is limited; the essential activity at an ICM is a process of digestion. When significant breakthroughs are made, they have to be made accessible and intelligible, simplified. In the span of one year, these breakthroughs can be counted on the fingers of one hand. At the same time, an indispensable ant-like labor is needed to incorporate the considerable number of auxiliary facts, which one would have great difficulty absorbing in their original form, but which are gradually simplified. What is important is not the amount of knowledge, but rather the conceptualization of the advances that have been made.

For about fifteen years, there has existed a very useful mathematical encyclopedia, published in Japan, which has the following virtue: if you are looking for some information, all you need to do is consult the index and refer to the right entry to find before you two or three pages of mathematical statements presented with sufficient clarity for a normal mathematician to have access to usable results.

It cannot be said that things are snowballing... I would argue to the contrary. In mathematics, the process of digestion is much slower than in physics, but the result is satisfying. This Japanese encyclopedia bears witness to that.[56]

---

[56] Alain Connes, André Lichnerowicz, Marcel Paul Schützenberger, *Triangle*

In the present context, what is perhaps most valuable to note are the points of convergence and divergence between Wittgenstein and Connes. Both are promoting a process of understanding through a slackening of pace, a process of "digestion." But whereas Wittgenstein would view this as the patient (yet heroic) work of the solitary philosophical mountain-climber, Connes views the process of "digestion" as occurring collectively within the community of mathematicians. And whereas Wittgenstein would suggest that the proliferation of mathematics be viewed as following upon a "loss of instinct," Connes would see the growth of the mathematical community– both in terms of number of participants and in terms of the production of results– as a *precondition* for the sort of understanding which currently coordinates the overall mathematical endeavor.

As Wittgenstein's view is most likely overly severe, so Connes' view is most likely overly sanguine, and it is probably no coincidence at all that the mathematical Platonist Connes is a firm inhabitant of the paradise that Cantor created and will surely never be driven from it. What strikes me as the most attractive aspect of Connes' position, and which does not depend on his Platonism in any obvious way, is the idea that "production" and "digestion" are integrated activities within the mathematical community, and that both are quintessentially *mathematical*. This avoids the rigid distinction between mathematics, grounded in truth, and ethics, grounded in praxis, which underlies Wittgenstein's orientation (as I read it), and which is surely one of the powerful sources of his ethical rigorism. What is perhaps most disturbing about Connes' view (although he does as much as he can to soften the blow) is the breathtakingly hierarchical view of the mathematical community which it implies. At the top are those very few associated with elite "breakthroughs," and the rest of the mathematical community constitutes a massive organism for the ant-like digestion and

---

of Thoughts, trans. Jennifer Gage (Providence: AMS, 2001), 100-01.

promotion of these orienting innovations, "countable on the fingers of one hand." This top-down picture is also a center-periphery picture, with the value of mathematical results gauged by their distance from the central mathematical breakthroughs of the day (not to mention the larger picture of how these contemporary innovations are folded in to the ongoing historical development of mathematics).

At root, however, there is an even more fundamental philosophical disagreement between Wittgenstein and Connes, which underpins their differences in conception of the growth of mathematics. This is the obvious disagreement between the two about whether the Platonist-Cantorian picture of mathematics constitutes an understanding of mathematics at all. Here the divide between Wittgenstein and Connes is unbrookable, and so we must return to a reconsideration of the philosophical status of the mathematical continuum in the conclusion to this chapter. In particular, I will focus on the connection between the mathematical concepts of number and continuum.

6.5 Conclusion (for this chapter and Track B): Number and Continuum

Wittgenstein sees a comparable problem in the cases of number and the continuum: in both cases there is a desire to specify an independent, abstract structure which reduces the logical structure of the respective domain under consideration to nonsense. In the case of number, this comes with the idea of numerical quantification leading to the idea of collecting the numbers together into a collection. This is a problem equally at the finite and at the infinite level. With respect to the continuum, the problem comes in attempting to specify structures in terms of extensional collections of point. Again, it matters not whether the collection of points is finite or infinite: the same logical problem arises in both cases. In fact, while the problem may be more tempting at the infinite level– since there we want to answer 'how many' questions on a par with finite cases– the problem may in fact be more *insidious* at the finite level. For here it is even *more* difficult to keep extensional collections and logical structures apart since we *can* in fact ask and answer 'how many' questions *in both cases*. ***Even at the finite level mathematics deals with structures, not collections.*** To put the point in Platonic terms, pure units are just as much a nonsensical fiction as the Cantorian continuum. In fact, it is no coincidence, on this view, that Cantor (at least at certain stages in his thinking) *builds up* his transfinite set theory with an original appeal to such pure, indiscernible units.[57]

Wittgenstein is thus calling into question a philosophical view of mathematical abstraction that extends historically from Plato to Cantor. In its own way, Cantor's Paradise is just an updating of the Platonic mathematicals, associated to the Platonic doctrine of forms. To continue to use Platonic terms: in this tradition, summoners are left behind, and we are "hauled towards being." The abandonment of scale, in particular, is the price paid for this view of independent, mathematical abstraction. For Wittgenstein, there is not a "relation" of the problem of the status of number as traditionally conceived to the problem of the continuum as traditionally conceived: logico-philosophically speaking, these problems are *one and the same*.

But suppose we *accept* Wittgenstein's critique and then go on to ask the question: what is the logical relation between number and continuum as structural systems? Here, Wittgenstein's princi-

---

[57]Compare the discussion in Michael Hallett, *Cantorian set theory and limitation of size* (Oxford: Clarendon, 1984), Sections 3.2 and 3.3, "Difficulties with the strange theory of 'ones'" and "The theory of 'ones' sensibly construed," 128-42.

ple of the relation of the general to the possibility of the particular provides a first clue. Whatever the answer to the question may be, it will have to be one which makes it possible to identify the logical links between the number system and particular numerical relations, and between the continuum system and particular relations of points.

In each of these cases, the relation must be one which *preserves relations of scale*. However, in the two cases, these relations of scale will differ significantly. This is because the number system is *intrinsically quantitative*: since each individual number stands for a quantity, the notion of quantity is ineliminable. *In the continuum system, a unit of measure must be specified.* There is no *intrinsic* specification of a unit of length. *I suggest that on Wittgenstein's view this is the **only** logical distinction that can be made between the two systems: the parafinite is the limiting conception of the structure of the continuum in its convergence with the conception of number.* We may formulate this as a slogan concerning the quantitative characterization of the parafinite, which I will therefore call the *principle of parafinite quantization*, or **PF 2**:

> **PF 2: *The parafinite is the quantitative reduction of the continuum.***

As I've formulated the slogan, it does *not* imply that the continuum and the number system differ *only* in terms of the absence or givenness of an intrinsic unit. But the slogan does imply that the parafinite can be viewed as a *reduction* of the continuum to this one "dimension." This reduction eliminates an intrinsically *geometric* content of the continuum. Wittgenstein's logical commitment to this reduction should send us back, I think, to a consideration of Wittgenstein's investigation of color and the notion of phenomenal space, but following that lead is beyond the bounds of this book.

The near relation of number and continuum as logical systems has dramatic implications, for it *relativizes* the notion of continuum in a way that I would like to try to specify in terms of an example. The example is not without its perils, since it is itself taken from an alternative form of set theory, and Wittgenstein's vigorous criticism of Cantorian set theory may apply here, too. It has the virtue, in any case, that it brings Wittgenstein into conjunction with an *alternative* set theoretic orientation that appears to have more in common with his view than the Platonism implied by Cantorian set theory ever could. Of course, in some fundamental sense it is the Platonism implied by taking an axiomatic presentation as a foundation for mathematics that is the deeper target of Wittgenstein's remarks, and not just the specifically Cantorian set theory. In a way, we are moving from the "tempting" to the "insidious," along the lines I have suggested above in the consideration of the extensional commitment to infinite versus finite number. My strategy will require first giving the example from alternative set theory, and then considering whether it can have any relevance once the set-theoretic nature of the example has been subjected to a Wittgensteinian critique.

Fernando Zalamea has made the very suggestive remark that in Vopěnka's alternative set-theory the natural numbers play the role of the continuum, being a super-multitudinous class.[58] Zalamea's discussion makes it clear that what is relevant for the relation between number and continuum in Vopěnka's "Alternative Set Theory" (AST) is that there are only two cardinalities: the cardinality of **N**, the "usual natural numbers," and the cardinality of the class **FN** of "finite" natural numbers, which are defined as those numbers for which every subclass is a set. The cardinality of **N** is "super-multitudinous" with respect to the cardinality of **FN**, and this allows it to play the role of continuum in Vopěnka's alternative set-theory. The situation could be usefully compared with Brouwer's intuitionism, in which there are also only two cardinal-

---

[58]Fernando Zalamea, *Peirce's Logic of Continuity* (Boston: Docent, 2012), 43.

ities: finite and countably infinite. Yet in Brouwer's intuitionism this second cardinality is far from exhausting the continuum: no countable collection can substitute for the appeal to the continuum in Brouwer's intuitionism. Zalamea points out that for Peirce the continuum is also super-multitudinous, but this involves arbitrarily high cardinalities, and so stands in this sense at furthest remove from Vopěnka's AST.[59]

At least on the face of it, Vopěnka's presentation of number and continuum in AST would bear some positive comparison with Wittgenstein's views as I have reconstructed them, and the ethical orientations implied by the two positions also seem roughly to align. Is there some sense in which we could see the super-multitudinousness of the natural numbers in AST as aligned with the idea that the continuum is lacking in a natural unit?

The idea is that no matter what unit we pick from among **FN** (the finite numbers), it will never establish a canonical unit for **N**: *no unit chosen from the finite numbers is adequate to serve as* ***the*** *unit for* ***N*** *because it can never be "large enough."* In this way, the supermultitudinous of **N** "models" the idea that there is no natural unit associated with the continuum. The "safeguarding" of **N** as continuum in AST is guaranteed by the *parafinitude* of **N**. Moreover, this is the *minimal* requisite for the protection of the supermultitudinous status of **N**, which is the *quantitative limit* of the conception of a (qualitative) continuum. The parafinitude is the minimal scale-breaking between the finite numbers and the parafinite numbers, which is required to guarantee the intrinsic unscaledness of the continuum: elements of **FN** are small and

---

[59]One should perhaps also note that the cardinality of **N** in Cantor's set theory is the only inaccessible cardinal *guaranteed* by the axioms of Cantorean set theory. Sometimes this notion is called *strongly inaccessible*, and sometimes the cardinality of **N** is ruled out by definition, in which case it is consistent to assume that there are no inaccessible cardinals in **ZFC** in this sense. See, e.g., Frank R. Drake, *Set Theory: An Introduction to Large Cardinals* (Amsterdam: North-Holland, 1974, 67-8.

elements of $\mathbf{N/FN}$ are large: here we find another movitation for the slogan that *the parafinite is the quantitative reduction of the continuum.*

Now we must go back and ask the question: does the axiomatic nature of this example invalidate it along the lines of Wittgenstein's critique?

On the one hand, Wittgenstein viewed the (axiomatic presentation of) the Cantorian continuum as a "gauze" because it obscured the fact that the notion of extensionally infinite collections didn't make any sense. But as we've seen, this was far from being the *most basic* problem, which was a circularity resulting from the fact that the definitions already implied what they sought to define. But, strangely enough, as we've also seen, Wittgenstein didn't focus on this more basic problem, even though it seems related to his general argument against a phenomenological language. Wittgenstein's focus was not on the circularity of Cantorian set theory, but on its *lack of perspicuity.* In particular, though not exclusively, this lack of perspicuity manifested itself in very specific ways having to do with the extensional picture of the continuum it suggested.

But does the Cantorian picture *imply* this extensional picture? If Wittgenstein is right, it *cannot*, at least coherently. But this is quite distinct from the suggestion that Cantorian set theory is *inconsistent.* Nowhere do we find Wittgenstein objecting to Cantorian set theory along the lines that it generates a contradiction, and he explicitly remarks that he is *not* saying that transfinite propositions are false. It really seems that the brunt of the objection to Cantorian set theory is that it doesn't *add* anything, and that there is something *morally* suspect because in fact it *seems* to.

With respect to the more particular point at hand, the problem with the Cantorian picture is that it seems to imply (albeit incoherently) that *the continuum is unscaled.* This violates Wittgenstein's dictum, as I've formulated it above, that the general must always have the particular as a possible case. If we take $\mathbf{N}$ as the continuum in Vopěnka's AST, there is no implication (whether incoherent or otherwise) that the continuum is unscaled. *In this respect it is a more adequate model of the continuum along Wittgensteinian lines* ***even if it is ultimately incoherent***. It is in this sense that I believe Vopěnka's AST is relevant for the consideration of Wittgenstein's critique of the Cantorian continuum, and at least some of this relevance is independent of whether Wittgenstein would find AST incoherent or not (I suspect he would).

This is not the end, but only the beginning, of a very long story. But for the moment it must serve as a provisional conclusion. In the Envoi, I turn in another direction, and one which leaves the logical finesse of Wittgenstein largely behind. If numbers are, as I remarked at the very beginning of this book, "the applier," then it is natural to look at an application. In the Envoi, I look at the application of numbers to the natural world and ask about the relevance of the parafinite in this venture.

Envoi: The Parafinite and the Natural World

As a sendoff to the patient reader who has made it this far in the drama of the parafinite's prehistory, in this final chapter I extend the examples provided by Galileo and Leibniz above to the application of the parafinite to the natural world. The development of the parafinite, whether understood as neither finite nor infinite (Galileo), or as infinite (Leibniz), emerged in conjunction with descriptions of various aspects of the natural world, such as the problem of material cohesion, and so this extension is not (forgiving the pun) unnatural. Two issues are deeply interwoven: the mathematical need for resolutive adequacy, i.e. the requirement that mathematical tools be sufficiently powerful to solve mathematical problems, and the need for descriptive adequacy, in particular for the description of the natural world. This interplay is related to, but not coextensive with, both Wheeler's complementarity of continuum and rigor, and Hardy's contrast between rigor and technique, both mentioned earlier in this volume.

In a contemporary setting, the investigation of the parafinite has been reinitiated in a large body of different work, whose variety I have barely begun to intimate. In this final chapter, I consider two types of recent characterization of the Leibnizian parafinitesimal, or indefinitely small: the description of Leibnizian infinitesimals by Abraham Robinson and Imre Lakatos in terms of non-standard analysis, and by Shaughan Lavine in terms of Jan Mycielski's analysis without actual infinity. The latter approach is more fruitful than the former, not least because it opens new possibilities for addressing questions about the relation between mathematical description and the world it describes, which at the end of the chapter I illustrate in the context of some issues in chaotic dynamics.

At the end of the chapter on Leibniz, I considered some remarks Leibniz made in a letter to Varignon which serve as a kind of master key for transit between the finite and the infinite in Leibniz's metaphysics. The passage articulates a form of the law of continuity, which allows Leibniz to assert that there is a kind of pre-established conformity between the finite and the infinite: the rules of the finite are found to succeed in the infinite and conversely. In particular, this means that we are allowed to manipulate (fictional) infinitesimals as if they were finite quantities and to treat small metaphysical beings as if they were infinitesimal.

In his essay, "The Metaphysics of the Calculus," Robinson discusses how his non-standard analysis may be used to model the appeal to infinitesimals in seventeenth-century calculus (particularly on the continent), citing extensively from de l'Hospital's *Analyse des infiniment petits pour l'intelligence des lignes courbes*.[1] Robinson recognizes that Leibniz's attitude toward infinitesimals was considerably more subtle than that of many of his correspon-

---

[1]G.F.A. de l'Hospital, *Analyse des infiniments petites pour l'intelligence des lignes courbes* (Paris: 1st ed. 1696, 2nd ed. 1715), cited by Robinson, "The Metaphysics of the Calculus," in *The Philosophy of Mathematics*, ed. Jaakko Hintikka (Oxford: Oxford, 1969), 153-63, 156. (It appears Robinson is citing from the 1715 edition, but this is not entirely clear to me.) For an introductory account of non-standard analysis see, for example, Herbert B. Enderton, *A Mathematical Introduction to Logic* (Orlando: Academic Press, 1972) 164-73. Robinson's full treatment of non-standard analysis relies on Henkin's non-standard interpretation of quantifiers. Stewart Shapiro, *Foundations without Foundationalism* (New York: Oxford University Press, 1991), 123-4 gives a helpful description of the interplay of first- and second-order logical ideas in the consideration of non-standard models of the real numbers.

dents, noting in particular Leibniz's criticism of l'Hospital's assertion of the reality of infinitesimals and his even stronger criticism of Fontenelle's commitment to them. Nonetheless, Robinson comments:

> ... Leibniz, like de l'Hospital after him, stated that two quantities may be accounted equal if they differ only by an amount which is infinitely small relative to them. And on the other hand, although he did not state this explicitly within his axiomatic framework, de l'Hospital, like Leibniz, assumed that the arithmetical laws which hold for finite quantities are equally valid for infinitesimals. It is evident, and was evident at the time, that these two assumptions cannot be accommodated simultaneously within a consistent framework. They were widely accepted nevertheless, and maintained themselves for a considerable length of time since it was found that their judicious and selective use was so very fruitful. However, Non-standard Analysis shows how a relatively slight modification leads to a consistent theory, or, at least, to a theory which is consistent relative to classical Mathematics.[2]

Robinson "slight modification" involves introducing a notion of equivalence which allows two numbers which differ only by a (non-standard) infinitesimal to be equivalent *without* their being identified as equal. In this way Robinson is able to provide a theory, consistent relative to classical mathematics, in which the conformity of infinitesimal quantities to finitary manipulation may be justified.

However, if we wish to consider how appropriate this mathematical formalism is for the description of Leibniz's own appeal to infinitesimals, we must look beyond what Robinson says. At

---

[2]Robinson, "Metaphysics," 158.

the end of this article Robinson does recognize that whereas Cantor's commitment to an actually existing infinity may be likened to l'Hospital's and Fontenelle's commitment to the reality of infinitesimals,[3] Leibniz's position is akin to Hilbert's original formalism, for Leibniz, like Hilbert, regarded infinitary entities as ideal, or fictitious, additions to concrete Mathematics.[4] But Robinson does little to indicate how non-standard analysis could be used to understand Leibniz's position in any detail, nor does he provide any further guidance in assessing the impact of his modification for the applicability of non-standard analysis to the description of historical positions like Leibniz's.

In these regards, Imre Lakatos' paper, "Cauchy and the continuum: the significance of non-standard analysis for the history and philosophy of mathematics,"[5] is more helpful. Lakatos uses Robinsonian non-standard analysis to uncover an exciting story of two rival theories of the calculus, revealed, however, at a surprisingly low degree of articulation.[6] The story pits these two rival theories of the calculus against each other in terms of their respective conceptions of the mathematical continuum: for a tradition of mathematicians stretching from Leibniz to Cauchy, a conception of the continuum as something enriched by non-standard values versus the Weierstrassian conception of the continuum (of which we find anticipations at least as early as Abel) in which such non-standard

---

[3]Robinson equally underlines Cantor's own antipathy to infinitesimals but calls his attempt to prove their impossibility on the basis of his set theory misguided (Robinson, "Metaphysics," 163). However, it is "misguided" with respect to *post*-Cantorian infinitesimals of the Robinsonian variety, not the pre-Cantorian variety to which Cantor objected.

[4]Robinson, "Metaphysics," 163.

[5]Imre Lakatos, "Cauchy and the continuum: the significance of non-standard analysis for the history and philosophy of mathematics," edited by J.P. Cleave, in Imre Lakatos, ed. John Worrall and Gregory Currie, *Philosophical Papers Volume 2: Mathematics, science and epistemology* (Cambridge: Cambridge University Press, 1978), 43-60.

[6]Lakatos, "Cauchy," 53.

values are excluded. Yet Lakatos cautions us against identifying the Leibnizian continuum with Robinsonian analysis:

> The decisive difference between Leibniz's and Robinson's theory of infinitesimals is exactly this: Robinson devises a *particular* non-standard analysis which is an elementary extension (in Tarski's sense) of real analysis, and where there are important bridges between the two analyses which make non-standard analysis testable. But this progress could not have been made before and without Weierstrass and Tarski.[7]

Neither is the Robinsonian continuum to be identified with Cauchy, for the same reason: "the Robinsonian construction... assumes a *particular* non-standard analysis of which Cauchy could not possibly have dreamed." Lakatos goes even further by claiming that "the continuity between Cauchy and Robinson is *much less* than the continuity between Cauchy, who enlisted infinitesimals, and Weierstrass, who did not."[8]

Key to Lakatos' analysis is his claim that it is the bridges between non-standard analysis and real analysis which make Robinsonian non-standard analysis testable. Itself lacking such a criterion of testability, Lakatos asserts, the Leibnizian continuum "was capable only of limited growth;" and it was "the heuristic potential of growth—and incredible explanatory power—of Weierstrass' theory that brought the downfall of infinitesimals."[9] Lakatos' point is primarily one about the *resolutive* power of the mathematical tradition associated with the employment of infinitesimals, as the following example makes clear:

> ...take as naive conjecture Cauchy's original thesis; take, say, Weierstrass's proof and Fourier's counterex-

---

[7]idem, 54.
[8]idem, 56.
[9]idem, 54.

amples as antithesis. The synthesis, the improved conjecture, will be the Weierstrass-Seidel theory at which
one arrives by finding the 'guilty lemma' in the informal
proof and incorporating it into the thesis.[10]

Lakatos juxtaposes his perspective to that of justificationist his-
toriography, which "would present the history of mathematics as
an accumulation of eternal truths." Such justificationism "leads
one either to date the history of mathematics from the date of the
last 'revolution in rigour', or to falsify the history of mathematics and reconstruct it in the up-to-date pattern."[11] Consequently,
Lakatos is at pains *not* to rewrite the history of mathematics in
Robinsonian terms. But it is difficult to see how Lakatos' appeal
to a Hegelian 'cunning of reason' which would turn each increase
in rigor into an increase in content is itself free of the charge of
writing history from the perspective of the victor. The point may
be made at a more subtle, but equally damning, methodological
level: even if Lakatos enlists non-standard analysis as a *tool* to
open up distinctions which the tradition articulated insufficiently,
the tool itself nonetheless opts in favor of the revolutionizing tradition of Weierstrass, and this must inevitably affect the morals
Lakatos draws. In particular, it underwrites his appeal to rigor as
an agent of historical development. I do not dispute that rigor has
been a driving agent in the history of mathematics, but in Lakatos'
hands rigor becomes (ironically) a focal, if not the only, motor of
historical progress. Although Lakatos will understand rigor on a
'critical' as opposed to a 'foundationalist' model, nonetheless his
joint commitments to critical rationalism and historical empiricism
drive him to accept the present historical vantage as de facto (albeit
always only provisionally) legitimate.[12]

---

[10]idem, 58-9.

[11]idem, 58.

[12]In these remarks I indirectly join a debate about the attitude of Lakatos
toward 20th century mathematics, particularly insofar as it is underwritten by

By virtue of his appeal to Robinsonian non-standard analysis, Lakatos has decided the debate between the Leibniz-Cauchy continuum and the Weierstrass continuum in favor of Weierstrass. For this reason (among others), Jan Mycielski's analysis without actual infinity serves as a more plausible point of departure for understanding Leibniz's approach to the mathematical continuum, since it need not rely on an anterior appeal to a conflicting conception of the continuum in order to present a model of Leibnizian analysis. Since, in fact, it is Shaughan Lavine who uses Mycielski's work to describe Leibniz's approach, I turn directly to his book, *Understanding the Infinite.*

Lavine represents the Leibnizian continuum as one in which our only access is to finitely small quantities which go proxy for infinitesimals in any given context. Depending on this context the smallness of these quantities will vary, and so we are not so much presented with a picture of the Leibnizian continuum as we are with a succession of approximations which suffice in different contexts. Yet this strategy of finite approximation is a virtue insofar as it is in keeping with Leibniz's dictum that the conformity of the finite to the infinite guarantees that infinitesimals may always be treated as small, but still finite, quantities. Thus, Mycielski/Lavine's approach mirrors Leibniz's methodological commitment to the law of continuity in a way that Robinson cannot: we actually work with finite levels but ones which, for the purposes of a particular problem, are *effectively* 'incomparable', i.e. "at different scales." Whereas

---

the support of axiomatic presentations. On this point the editorial comments appended to *Proofs and Refutations* seem likely to serve as an initial point of engagement. I am grateful to David Corfield for circulating unpublished material he has written on Lakatos at the Summer 2001 NEH Seminar, "Proofs and Refutations in Mathematics Today," co-directed by David Corfield and Colin McLarty at Case-Western Reserve University, and to the participants in this seminar for helpful conversations on these and related themes. These remarks also take up issues discussed in *The Pace of Modernity: Reading With Blumenberg*, 191.

Robinson's non-standard analysis may reasonably claim to model the incommensurability of levels, in no way does it capture the sense that we are to think of Leibniz's infinitesimals operationally as incomparably small but nonetheless finite. That is, Leibniz's position is that in any individual case the fictional infinitesimals may *actually* be taken as sufficiently small finite quantities. This is guaranteed by the ideal conformity between the finite and the infinite, underwritten by appeal to the law of continuity. Incidentally, it also suggests that while, in principle, the Leibnizian indefinite can be understood as potential or actual as you please, in mathematical *practice* it is understood as potentially infinite/infinitesimal and actually finite. Since, conceptually, *both* the potential and the actual are enlisted, the indefinite must be capable of being understood as either potential or actual "as you please."

It is this set of features that is central to Leibniz's approach to the calculus insofar as his treatment of the natural world is concerned. As Domenico Bertoloni Meli has pointed out, in conjunction with the conformity of the finite and the infinite, Leibniz's approach to the description of physical processes is to treat the mathematical description in terms which represent the physical process by picturing a finite number of physical events. Thus, for example, the trajectory of a body under the influence of a series of impacts is described mathematically in terms of a polygonal approximation to that curve. The curve itself is then thought of as an infinite polygonal trajectory in which each of the steps is infinitesimal. In this way,

> The polygonal curve, however, corresponds in principle to physical actions in a way that the continuous curve does not. By correspondence 'in principle' I mean that mathematics mirrors the physical laws involved, rather than the infinitesimal details of the trajectory of the body.[13]

---

[13]Domenico Bertoloni Meli, *Equivalence and Priority: Newton versus Leib-*

Bertoloni Meli's point is key to understanding how Leibniz's mathematics is geared to the description of nature: not by giving us a picture of how bodies actually (in any sense) move, but by providing a mathematical structure which embodies the physical laws at some (ever refining) finite level of approximation.

In conjunction with the above, we should also consider a point which Meli makes concerning Leibniz's treatment of dimensional inhomogeneity. This point needs to be addressed in responding to the potential criticism that the Mycielski/Lavine model puts the derivative at the heart of the calculus, whereas Leibniz's approach proceeds in terms of a focus on the concept of the differential. Since, Meli notes, in coping with dimensional inhomogeneity, such expressions often involved infinitesimals, one can say that the differential proportion was the primary tool in the mathematical investigation of nature. Leibniz's approach is perfectly accommodated to this since his extended conception of quantity allows him to take differences (respectively) of infinite, finite and infinitesimal quantities. But the focus on the differential is not, as in both the Robinsonian and Mycielskian reconstructions of analysis, a waystation on the road toward the definition of the derivative; rather, it is focal *in the absence of* an operant conception of the derivative. As I've described above, Bos has shown how the focus on the differential allows us to see both the strengths *and* the weaknesses of the Leibnizian calculus: on the one hand its capacity to work naturally with changes of variables even at the infinitesimal level, on the other its incapacity to provide a foundational justification for the requisite rules of calculation.[14] One of the results of Bos' presentation of the Leibnizian calculus is that we see that the foun-

niz, including Leibniz's Unpublished Manuscripts on the **Principia** (Oxford: Clarendon, 1993), 83.

[14]Meli asserts that Leibniz had no conception of function, but I don't think this is right, as Dietrich Mahnke has argued in "Die Enstehung des Funktionsbegriffes," *Kantstudien* **31** (1926), 426-28. Still, Leibniz certainly did not make such a conception focal. Consequently, the force of Meli's point stands.

dations of the calculus, and in particular the attempt to justify the application of higher-order infinitesimals, required the (arbitrary) choice of a preferred variable which could remain fixed for purposes of foundational justification. But, as Bos points out, this has the negative pragmatic consequence that the choice of progression of variables, in terms of which the Leibnizian calculus exhibits its intrinsic power, is hindered by such an arbitrary fixation. As Bos also shows, Leibniz himself was aware of how fixing an independent variable could be used to justify first-order differentials, and to this extent he anticipated the development of the calculus in the direction of a functional characterization. Further, despite the fact that Leibniz did not successfully extend this (for Leibniz, presumably rather awkward) approach to the justification of higher order infinitesimals, Bos shows how this can be done, but again at the cost of constraining the progression of variables by specifying a single fixed, independent variable. Both Leibniz's justification in the first-order case and Bos' analogue in the higher-order case rely on a form of Leibniz's principle of continuity, whereby "if any continuous transition is proposed terminating in a certain limit, then it is possible to form a general reasoning, which covers also the final limit" (cited in DLC, 56).

Insofar as it is grounded in, or at least compatible with, a functionally-driven approach to analysis, the Mycielski/Lavine reconstruction is potentially subject to the charge of reconstructive anachronism in a manner similar to Robinson's reconstruction above. However I am not arguing that the Mycielski/Lavine reconstruction (or indeed any reconstruction) can be entirely faithful to the material it seeks to reconstruct, as is practically tautological in the notion of 'reconstruction'. Instead, I am only claiming that the Mycielski/Lavine picture is not in fundamental conflict with the Leibnizian conception of the continuum, as I claimed the Robinsonian approach to be. In order to see this in more detail, consider the respective definitions of the derivative in the Robinsonian and Mycielskian reconstructions. In Robinson's analysis, it is a necessary and sufficient condition that c is the derivative of $f(x)$ at $x = x_0$ if c is the standard part of

$$\frac{f(x) - f(x_0)}{x - x_0}$$

for all $x \neq x_0$ in the monad of x_0.[15] Here the result on derivatives depends on the prior definition of the monad of any (standard or non-standard) real number, which consists of all those points which are infinitely close to the original number. Logically speaking, then, we need to know the value of the difference quotient for all values infinitely close to x_0 in order to determine the derivative. Note that from a post-Weierstrassian perspective this is not so much a disadvantage as it is an advantage: it leaves the door open for all sorts of functional behavior that would be pathological and undreamed of in the Leibniz/Cauchy universe. In Mycielski's analysis without actual infinity, on the other hand, the derivative d of a function f at a value a at a particular level is defined in terms of one specific quotient of differentials:

$$d =_0 \frac{d_1 f}{d_1 x} = \frac{f(a + \zeta) - f(a)}{\zeta},$$

where ω is the largest number at this level and where $\zeta = 1/\omega$.[16] The equality "mod-zero" means that we take the answer we get on the right hand side and drop any "parafinitesimal" quantity remaining from the chosen level of refinement, retaining only the "finite" part. Once again, the Mycielskian approach captures the fact that we should not think of the Leibnizian continuum as consisting of a pre-existing supply of individual points.

But, in fact, the Mycielskian approach is illuminating in an even further regard, for it shows why, without fixing a particular independent variable, i.e. a particular standard infinitesimal, on the

---

[15] Robinson, "Metaphysics of the Calculus," 155.
[16] Lavine, *Understanding the Infinite*, 281-2.

face of it we should have no reason to believe that higher derivatives of a (differentiable) function are well-defined: why should we expect the ratios obtained on various levels to give us the value of the higher derivatives? That is, why should we expect them all to agree? That they do (for a differentiable function!) is only guaranteed in the Mycielski/Lavine picture by the fact that the various infinite numbers we introduce are taken to be indiscernibles.[17] This would be true of Leibnizian infinitesimals too since they are abstract, but not of the nested physical levels which the finite mathematical proxies literally describe! In Leibniz's own case it is only the application of the principle of continuity, which guarantees that quantities will behave in the (fictional) infinite as they do in the finite, that thereby guarantees mathematical indiscernibility.

This does not exhaust the work required of the principle of continuity, either, for as Robinson has pointed out, Leibniz's own calculus is, in fact, inconsistent. In order to overcome this problem, Mycielski and Lavine have introduced adequalities (i.e. equalities "mod-zero") and indiscernibles, which are necessary to prevent their logical reconstruction from running afoul of inconsistencies which both the authors recognize as intrinsic to Leibniz's approach. But Leibniz handles the problem of adequalities with reference to the law of continuity: the application of this principle justifies our ignoring infinitesimal quantities in the limit. And, to the extent that he indicates a path for addressing it, the problem of rendering the various orders of infinitesimals compatible is handled by fixing a standard unit of measure. Leibniz is not completely successful either on the front of the law of continuity or on the front of fixing a standard infinitesimal unit: we might say that in both cases Tarski wins out over Leibniz. But in Mycielski/Lavine's reconstruction *Weierstrass does not*: in the Mycielski/Lavine approach,

---

[17]It would be interesting to compare this hypothesis of indiscernibility with assumptions about self-scaling, which suggests a potential connection with fractal geometry.

we supply points in the continuum *as it becomes necessary*.

Further, unlike Robinsonian non-standard analysis, for which there is a global result showing that all theorems of standard analysis having a proof in non-standard analysis have a standard proof analogue, in Mycielski's analysis without actual infinity, it appears that we obtain results with no classical analogue. This leaves open the possibility that Mycielski's calculus may in this sense be richer than classical analysis. Bertoloni Meli has stressed the methodological appeal of Leibniz's calculus as it is used to describe the natural world. Does it have anything to tell us so far as contemporary physical modeling is concerned? Although it is not incumbent on the understanding of history to yield such fruits, the specific nature of Mycielski's reconstruction gives us reason to think that in this particular case it might. In particular, Mycielski's finitization of portions of his theory allows us to prove results up to small probability that don't have any (obvious) analogues (with proofs) in the infinitary case. Recently, several physicists have renewed Emile Borel's recommendation that an axiom be introduced for physical theories which states that if probabilities are sufficiently small then they can be taken as negligible and so identified with zero. Indeed, a program along these lines has been described in some detail for the foundations of quantum mechanics by Roland Omnes.[18] Adoption of such an attitude would provide a warrant for pursuing such results of Mycielski type.

Failing this, however, we may still ask about the appropriate-

---

[18]See, in particular, Roland Omnes, *Understanding Quantum Mechanics* (Princeton: Princeton, 1999), and in particular his Emile Borel's work in *Valeur pratique et philosophie des probabilités* (Paris: Gauthier-Villars, 1937), at pp. 84, 237, 239, 291n (here Omnes credits Pierre Cartier, and it's worth remembering Cartier's connections to the French school of non-standard analysis, which has a very different orientation than Robinson's school, in general. See the discussion in Jean-Michel Salanskis, *Le Constructivisme non-standard* (Paris: Presses Universitaires du Septentrion, 1999). I take these points up in the discussion of J. L. McCauley's work below.

ness of continuum modeling for the description of the physical world. Here advances in chaos dynamics point to difficulties with continuum modeling, in particular because these investigations cast some doubt on the assumption that mathematical conditions obtaining with probability one are indeed representative of physically generic conditions. Building on the work of Borel and many others, J.L. McCauley has argued that there are physical grounds for believing both that (metaphorically speaking) nature does not necessarily make a random draw from the continuum and that finitary modeling procedures, constrained only by the availability of computer time, are in fact perfectly adequate for the modeling of mathematical behavior that is as a matter of fact most often presented in a continuum context.[19] McCauley in fact goes further to stress the great advantages of such finitary modeling over traditional continuum-based approaches, and at a point which directly links the two issues mentioned above (CDF, 189): the provision of a finitary algorithm for the description of a chaotic process simultaneously provides us with the algorithm which induces the most natural coarse-graining of phase space. Here McCauley is proceeding in terms of the same sort of regulative principle which Bertoloni Meli isolates as governing Leibniz's approach: the mathematical approach is chosen to optimize the fit between the mathematical description and the physical laws governing the process described. To explain this in the setting of chaotic dynamics would require a discussion of the foundations of statistical mechanics, in particular, which is well beyond the bounds of this book, but the regulative principle is in both cases the same.

In the rest of this envoi I will focus on the way in which natural partitioning may be used to break both mathematical and physical symmetry. I will suggest that, well beyond there being good

---

[19] J. L. McCauley, *Chaos, Dynamics and Fractals: An Algorithmic Approach to Deterministic Chaos* (Cambridge: Cambridge, 1993), 84; hereafter cited in text as CDF.

mathematical reasons for doing this, this is in fact a *mathematical necessity*, and one that plays a particularly significant role when we seek to apply a particular mathematical description to the physical world. This leads to my fourth and final slogan for the parafinite. Since it concerns the way the parafinite identifies the asymmetry embodied in the finite/infinite distinction, I call it the *principle of parafinite asymmetrization*, or **PF 3**:

> **PF 3:** *The parafinite is the broken symmetry between the (de)finite and the in(de)finite.*

Every time we choose a finite number from the natural numbers, or select a unit of measure for the continuum, we break the symmetry within the natural numbers or the continuum and establish a distinction between the finite and the infinite. In the first case, the numbers less than the one I have chosen are finite, and the numbers more than the chosen one are infinite. In the second case, the unit of measure chosen is finite and the reduplication of this unit of measure within the continuum is infinite. The analogy between the two cases suggests that in thinking about the continuum, it is the relation between a bounded line segment and the unbounded continuum which is primary, and not the relation between "a point" and "the line." The Greek emphasis on lines as segments bounded on both ends is thus closer to this analogy than the Cantorian construction of "the continuum" out of "points," suggesting that at this level of conceptual depth there is some alignment between the historical-genetic and conceptual aspects of the issue. But for the Greeks, the fixing of a bounded line segment was the "wresting," if you will, of a definiteness from something indefinite, unbounded. Hence I have formulated this final principle of the parafinite in a way which applies both to the finite/infinite and to the definite/indefinite distinction.

In some sense, the recognition that choosing a number, or a unit, "breaks" the symmetry of the numbers or the continuum is an obvious one. What is not obvious is why (or indeed even if) this recognition should be fundamental for our understanding of quantity or extension. Indeed, it is with the goal of providing some historical-conceptual motivation for the fundamental nature of this recognition that this volume has been composed. If the motivation is found persuasive, what remains is the ambitious task of mapping out the ramifications this history portends. This major task requires a collective investigation, well beyond the bounds of any individual to fulfill. The history I've presented suggests that *many* preliminary steps in this direction have *already* been taken, and that part of the work needed is to recognize them *as* steps. McCauley's work counts as such, so let me begin by drawing on one of his examples.

McCauley points out that algebraically we may operate with variable symbols in the abstract, and in fact in the interest of ease we often do. But if we want to *compute* (trajectories, say) then we have to introduce explicit representations (e.g., symbol strings with respect to a given base). In fact, McCauley's point can, and should, be generalized to the context of using mathematical structures to characterize physical (or, indeed, any other, including mathematical) systems: just as an abstract mathematical structure must be couched in a particular calculus, so our application of mathematics to the descriptions of nature (or other systems, including mathematical ones) should, on the one hand, preserve maximum flexibility by remaining basis-free (following Bertoloni Meli's point about embodying physical theories). On the other hand they should also allow themselves to be adapted approximatively to whatever count in the relevant context as natural frames (and finite approximation, in particular). When we tie ourselves down to a particular frame for *foundational* reasons, this may lead the mathematical theory to become unperspicuous in its application to the description of nature. Here one should recall the tradeoffs associated with the specification of one fixed progression of variables in the Leibnizian calculus.

This lack of perspicuity is arguably exactly what occurs when we use our well grounded continuum-based approach to analysis and apply it to certain physical theories: what we have adopted because it gives us a good mathematical foundation encourages us to identify measure one mathematical behavior with physically generic behavior. *But this move is not warranted, in general, because we have no reason to believe that the way that generic physical behavior is selected has anything to do with the way our mathematical analysis has become fixed.* In fact, there is good reason to think that generic physical behavior is often selected in terms of symmetry breaking via privileged frames or partitions. This point can also be understood in a purely mathematical context by looking at how we represent particular real numbers. Here the symmetry is broken, in particular, by the choice of a particular base for numerical representation. Leibniz sets up the calculus in exactly the same way if we think of the finite proxies for the fictional infinitesimals as breaking a symmetry with respect to *scale*.[20] Consequently it should come as no great surprise that it is exactly in physics where scaling behavior is involved that the problems with continuum-based mathematical descriptions of nature become most readily apparent. As McCauley shows at length, when dynamical behavior is associated with a positive Lyapunov exponent (roughly: positive entropy) we are faced with a situation in which the least significant digits of an initial condition are amplified, often dramatically, and we experience what has come to be identified as the butterfly effect: small changes in initial conditions leading to exponentially divergent trajectories which may quickly lose all significant correlation.

McCauley contrasts two divergent (!) repsonses to the specific challenges placed on modeling by the decay of such correla-

---

[20]It then seems that the application of the law of continuity *across* scales amounts to an assertion of ideal rescaling.

tions. The standard attitude to is to carry out the numerics by using floating-point or fixed-precision real arithmetic on a digital computer, and then let measure theory be our guide as to what we should expect as computer output (CDF, 280). In particular we appeal to measure theory in support of a belief that the output of such approximate calculation should exhibit mathematically generic, i.e. measure one, behavior. But, McCauley argues, the appropriateness of drawing on probability theory in the interpretation of computations depends both on (i) the availability of the continuum, and (ii) the possibility to make a *truly random choice*. In general, McCauley marshals arguments bearing on physical appropriateness with regard to the first point, focusing on the fact that so far as *actual* computations are concerned we are thoroughly restricted to the finite domain and that, even so far as computability is concerned we are restricted to a countable number of such functions. Regarding the second point, McCauley leans on the fact that statistical pseudo-randomness can be generated by an entirely deterministic procedure, so that we have no positive reason for thinking that the notion of a truly random choice is even a coherent one.[21] Indeed, given that we only possess a single independent algorithm for generating a normal statistical distribution within a number string, an approach which attends to the foundational role of computation in the physical modeling involved will be unlikely to adopt the second assumption unreflectively. Not surprisingly, McCauley does not.[22]

The second approach, which McCauley favors, is, in a well-defined mathematical sense, inverse to the one described above: "in

---

[21]This may be the most readily available argument to support the second point, but philosophically I do not think it is very strong. What McCauley goes on to say, as I report in the next sentence, strikes me as more significant. But ultimately the problem deserves much fuller treatment than McCauley gives it or than I can give it here.

[22]Compare Thomas P. Weissert, *The Genesis of Simulation Dynamics: Pursuing the Fermi-Pasta-Ulam Problem* (New York: Springer, 1997).

direct contrast with the advice of most other texts on deterministic chaos, we have not advocated the forward iteration of a chaotic map on a digital computer (in the floating point mode) as a method for discovering the statistical distributions of determinstic chaos." (CDF, 280) Instead, McCauley advocates studying deterministic chaos by looking at the natural partitioning of the space generated by the inverse image of the map under investigation. What emerges is an approach to deterministic chaos that is in conformity with Mycielski's reconstruction of analysis in important regards. With progressive iterations of the inverse map we generate finer and finer partitions of the space, and each of the elements of these refining partitions remain together for longer and longer extents of their trajectories. We may thus think of these refining partitions as grouping trajectories together with respect to the original partitioning of the space, or we may think of the progressive refining of partitions as a progressively finer coarse-graining of the space, and one naturally fit to the dynamics of trajectories. If the original partition is chosen to optimize information about the dynamics of the system this information will be carried, and indeed refined, through the refinement of this coarse-graining. Instead, then, of studying the statistics of forward 'test-trajectories' which we can only calculate approximately and where it remains unclear whether our information bears on the dynamics of the system or is rather an artifact of our computer's rounding algorithm, we accumulate exact information about the (finite) statistics of (finite) trajectories through the space. From a measure-theoretic approach these statistics are transient since we can always construct mathematical examples in which radically different finite statistical information converges on the same measure-theoretic information in the limit (and vice versa).[23] The alternative approach more naturally em-

---

[23]This recognition served as one of the two starting points for my thinking about the parafinite as a mathematics graduate student at Wesleyan University. The other involved recognizing that the functional relations among depen-

bodies the statistical properties we seek to study as well as naturally focusing attention on finite statistics, which in any case is all we can expect to model physically. In some sense it is even more mathematically natural along the funadmental conceptual lines of measure-theory, since the measurability of a function is specified in terms of the inverse images of the function in question.

The "inverse" approach in terms of finite statistics shares the virtue attributed by Bertoloni Meli to the Leibnizian calculus: since in this case it is the structure of statistical correlations which we seek to describe, the mathematics mirrors the statistical laws involved, rather than the details of the trajectories of individual orbits. In this sense it would be appropriate, I believe, to describe this approach as structural, and it is with some remarks on structure and representation that I will now conclude this labyrinthine tour through shadow-castings of the parafinite.

As mentioned above, McCauley highlights the tension between variable symbols which we manipulate (structurally) in the abstract and explicit representations (symbol strings with respect to a given base, say) which we use to calculate. Internal to the treatment of deterministic chaos this tension is perhaps most important in terms of the tension between topological structures, which are invariant in a strong sense, and measure-theoretic and metric structures, which lack the same (desirable) level of invariance. It is not unfair, I think, to say that McCauley's campaign against measure theory in general is driven by his desire to purge the study of determistic chaos of such variant, representation-dependent methods. Whether analogous measure-theoretic invariants exists is at best unclear, and independently of this issue, McCauley's approach possesses all the virtues of natural coarse-grainings. It may ultimately be that the prooblem with measure theoretic approaches is rather that they don't give us any geometric hold on the measure-algebra structures,

---

dent quantities chosen in the course of a mathematical proof cannot in general be bounded so far as the growth-rate of dependent quantities is concerned.

at least independently of a scaling geometry. This is why fractal geometry is useful to "flesh out" measure-algebra structure.[24]

The turn to an invariant approach is a familiar process in the history of applied mathematics, perhaps best seen in this century in the shift from matrix and tensor methods to methods associated with differential forms. In the study of deterministic chaos this drive takes the form of an attempt to formulate dynamic properties which are universal, i.e. those for which the details of the initial conditions are *irrelevant*, so that they depend *only* on the natural partitioning of the space and hence may be studied exclusively in terms of (invariant) symbolic dynamics (CDF, 241). Once again, the grounding is in methodological principles which are empirically driven: the motivation for starting with a symbol sequence rather than with an initial condition is that the former *may* be extracted from certain experimental time series while the latter cannot be obtained in any direct way from experiment. This does not relieve the physical modeling of difficulty, but does focus it upon the issue which is arguably of most relevance because of its methodological transparency. The difficulty– and it is a nontrivial one– is that one must have a way of extracting the natural partitioning unambiguously from the data in order to carry out the procedure. In the penultimate chapter of the book McCauley gives an example of the difficulties faced in the particular case of modeling eddy-cascades

---

[24]One of the most interesting and potentially fruitful programs in this regard is that of Michel Lapidus, who has developed a sophisticated approach to fractal geometry using a novel definition of complex dimensions. See, in particular, M. L. Lapidus and M. van Frankenhuysen, *Fractal Geometry, Complex Dimensions and Zeta Functions: Geometry and spectra of fractal strings* (New York: Springer, 2006), and for more general implications, M. L. Lapidus, *In Search of the Riemann Zeros: Strings, Fractal Membranes and Noncommutative Spacetimes* (Providence: AMS, 2008). In conversation, Benoît Mandelbrot indicated that his multifractal analysis was intended as a replacement for measure-algebraic ergodic theory, which he viewed as of very limited value for physical analysis. It would seem fair to characterize Lapidus' program as following instead a "both-and" approach.

with low Reynolds number, and the challenges encountered are sobering.

The moral, then, is not to simplify our lives, and so, pragmatically speaking, the methodological virtues practiced by Leibniz, Mycielski, and McCauley may not yield the quickest fruits. There are virtues associated with the perspicuity of particular rigid representations that this chapter has left entirely unexplored. In other cases, a rigid representation may be the only one currently feasible. The further weighing of the relative virtues of such theories is an even more difficult problem, and one perhaps best left to the historical and conceptual analysis of individual cases until a greater appreciation of the taxonomy of such virtues is attained. In this final envoi it has been my goal only to point to some connections between the parafinite and the virtues of certain approaches to mathematical modeling.

Appendix A

Creating Larger and Larger Numbers with Functions

In this appendix, I will show something which essentially gives the answer "yes" to this question asked previously: are there functions which produce numbers even larger than the functions on the list of "arrow" functions? I will show that we can make a function out of this very list so that, for large enough inputs, the outputs are bigger than for any single function in the list you can pick. You will probably think there is something tricky about this (in the pejorative sense), but the principle turns out to be very important for constructing wildly large numbers. So bear with me a minute—and if the argument is too complicated to digest on first reading, you can always come back again later.

To simplify matters, I'm going to construct a function with only one input; I'll call it $G(x)$. To do this, I have to say what the output is for any given input value x. The rule I will adopt is very simple:

$$G(x) = g_x(x, x).$$

So, for example, $G(2) = g_2(2, 2) = 2 \uparrow\uparrow 2 = 2^2 = 4; G(5) = g_5(5) = 5 \uparrow\uparrow\uparrow\uparrow\uparrow 5$, which is already well too large to calculate explicitly. Now here's the claim: *for large enough values of x*, $G(x)$ is bigger than $g_n(x, x)$. If you think about the way $G(x)$ is defined, it's clear that this is true as soon as x is bigger than n. For by definition, $G(x) = g_x(x, x)$, and as soon as x is bigger than n, $g_x(x, x)$ is (much) bigger than $g_n(x, x)$. Think, for example, of how much larger $G(5) = g_5(5, 5) = 5 \uparrow\uparrow\uparrow\uparrow\uparrow 5$ is than $g_2(5, 5) = 5 \uparrow\uparrow 5$, for example (taking $x = 5$ and $n = 2$). This procedure is called *diagonalization*, since we define $G(x)$ by picking the values of the functions off the diagonal of the table which arranges the values of the original g_n functions:

$$
\begin{array}{ccccc}
G(1) & g_1(2,2) & g_1(3,3) & g_1(4,4) & g_1(5,5) \\
g_2(1,1) & G(2) & g_2(3,3) & g_2(4,4) & g_2(5,5) \\
g_3(1,1) & g_3(2,2) & G(3) & g_3(4,4) & g_3(5,5) \\
g_4(1,1) & g_4(2,2) & g_4(3,3) & G(4) & g_4(5,5) \\
g_5(1,1) & g_5(2,2) & g_5(3,3) & g_5(4,4) & G(5)
\end{array}
$$

and so on (I have only listed entries here where $x = y$, since those are the only relevant ones in defining the function $G(x)$). By diagonalization, we build a function which eventually (that is, for large enough input values) outstrips every function in the list. We can then reboot the process of iteration described in the chapter above to create functions growing even faster than $G(x)$, and on and on. In principle there's no end to this process, but if you try to implement it, say on a computer, it very rapidly becomes unfeasible.

Appendix B

Rolling Polygons

You may have been wondering about the details of the upper part of the diagram presented in Figure 3.1. As I've already described, Galileo uses this polygon to motivate his solution to the problem in terms of the interposition of gaps, but let's look at this in a bit more detail. The polygonal case shows how a figure with a finite number of sides solves the problem through the interposition of finitely many finite-sized gaps in the trajectories of interior polygons. Galileo then goes on to motivate the solution for the circular case by appealing to a kind of limit of the finite case, in which the continuum is resolved into indivisibles in one fell swoop, as discussed above.

When we roll the hexagon $ABCDEF$ along the line extending from A to S, the sides of the polygon will cover the length of the segment from A to S, while the sides of the smaller polygon $HIKLMN$ will leave gaps along the line HT as the smaller polygon rolls along. In the first step, when the side BC rolls down onto the bottom line AS in the position BQ, IK will roll past the segment IO and land in position OP on the line HT, leaving a finite gap in the position IO. Next time, when CD rolls down and lands in position QX, KL will roll over PY, leaving a gap there, and land in the position YZ. This explains how the smaller polygon can keep up with the larger one: it does so by leaving many little (but still finite) gaps in between the places where sides land on the line HT. But once we are dealing with circles there are no little gaps, and the smaller inscribed circles trace out the same length as the large, exterior one. In Figure 3.1, the circle with radius AB traces out the length from B to F while the smaller circle with radius AC traces out the same length from C to E. Galileo will resolve this puzzle by appealing to non-finite, indivisible voids.

Galileo's proposed solution, then, involves a qualified denial that there are no gaps in the trajectory of these smaller circles. Reasoning by analogy with the case of the polygon, in which finite gaps are interposed, he will insist that when the smaller circles are laid out along the corresponding line infinitely many indivisible voids (GO, 72) are interposed! These voids are not finite gaps such as we find in the case of the polygon, and so in this case the gaps have no measure. Instead they are non-quantitative, but infinitely many. In the end, the interposition of these infinitely many indivisible vacua will stretch the line out to the length of the circumference of the exterior circle.

To see this in terms of the comparison between the polygon and the circle, we need to run the movie backwards, so to speak. Think about taking the line AS and wrapping it up into a polygon like $ABCDEF$. In order to do that, I would have to mark the point B and then bend the line AS to make the angle ABC. Then I would have to mark the point Q and bend the line again to make the angle BCD, and so on. All in all, I would have to mark six points to serve as the vertices A, B, C, D, E and F. Thus we have made a finite number of divisions (in this case, six) of the line AS at the points B, Q, X, and so on. Now, by analogy, imagine rolling the line BF up into a circle. This would involve making a mark at every point along the line for the entire circumference of the circle, and so in this sense you would have to resolve the relevant portion of BF (corresponding to the length of the circle's circumference)

240

down into every single point along this line! So when the circle with radius AB is rolled up out of the line BF, the continuum BF is resolved into an infinity of points at one fell swoop.

From this we see that by dividing a line finitely we cannot rearrange it into something larger, yet by dividing it infinitely we can put it back together into something so much larger as we please.[1]

---

[1]This should be compared with the so-called Banach-Tarski paradox. Consideration of Galileo's example motivates the perspective that what is intrinsically involved in the expanding spheres of the Banach-Tarski paradox is the role that the axiom of choice plays in allowing us to establish the sort of decompositions of the real numbers which are analogous (in the relevant sense which, of course, remains to be specified) to Galileo's infinite decompositions. On the Banach-Tarski paradox, see Stan Wagon, *The Banach-Tarski Paradox* (Cambridge: Cambridge University Press, 1985). As Knobloch points out, the indivisible voids are *unmeasurable* ("Galileo and Leibniz," 93), so that in both cases expansive rearrangement rests on unmeasurable entities (in the respective senses of 'unmeasurable', of course).

Appendix C

A Proof for the Impossibility of Infinitesimals

The Leibniz scholar André Robinet has located a manuscript passage which he believes reproduces the proof to which Leibniz refers in his 1702 letter to Varignon, and Enrico Pasini has dated this passage from Leibniz's first Hannover residency, which is to say in the years after he developed the calculus in Paris (1672-1676). The passage is crossed out, indicating that Leibniz ultimately abandoned the proof, and apparently the idea is not taken up again elsewhere. Given that Leibniz ultimately rejects the proof, we should not expect too much from this passage. But inspecting it helps demonstrate the issues involved in Leibniz's ongoing development of his account of the indefinite as infinite, so I want to present the argument here.[1]

---

[1] The material in this appendix draws on O. Bradely Bassler, "An enticing (im)possibility: infinitesimals, differentials, and the Leibnizian calculus," in Ursula Goldenbaum and Douglas Jesseph, eds., *Infinitesimal Differences: Controversies between Leibniz and his Contemporaries*, (Berlin: Walter de Gruyter, 2008), 135-51.

Let me begin by attempting to put my basic understanding of the argument into English prose. To help with this, I include here a diagram (based on one Leibniz gives) illustrating the relevant quantities for Leibniz's proof.

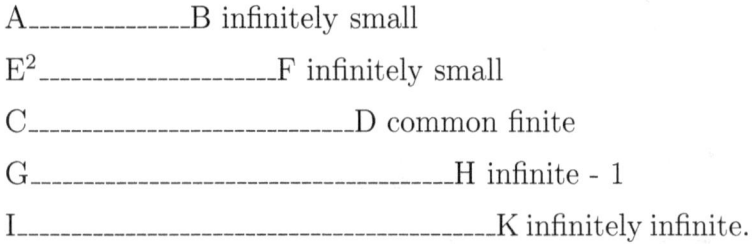

A_____B infinitely small

E^2_____F infinitely small

C_____D common finite

G_____H infinite - 1

I_____K infinitely infinite.

The argument works as a proof by contradiction. That is, we begin by assuming the opposite of what we want to prove, and then we derive a contradiction from it. Since the opposite implies a contradiction, we conclude that the opposite of the opposite, i.e. the original claim, must be true. So, since Leibniz wants to prove that infinitely small magnitudes are impossible, he begins by assuming the existence of an infinitely small straight line AB and seeks to derive a contradiction. The fact that we are dealing with geometric magnitudes here will be of utmost importance. Leibniz juxtaposes to this infinitely small line AB a normal finite line segment CD and tells us to take a mean proportional between the two segments, which he calls EF. Here a mean proportional is a (geometric) magnitude satisfying the equation:

$$\frac{AB}{EF} = \frac{EF}{CD}.$$

EF is mean in the sense that it stands as a middle term between AB and CD: AB stands in proportion to EF as EF stands in proportion to CD; even though these are geometric magnitudes,

---

[2]Here I am reading 'E' for Robinet's 'b', following Enrico Pasini's transcription.

we can read these proportions as fractions. If EF is a mean proportional, then in particular EF must be greater in length than AB and less than CD. But why should a mean proportional exist here? Leibniz does not make this clear. This is particularly troubling given that the magnitude AB is infinitesimal in length and the magnitude CD is finite in length. I will return to this point later; for now, along with Leibniz, I simply assume the existence of such a mean proportional. If EF is to be a mean proportional between AB and CD, it would have to be either infinitely small (although greater than AB) or finite in length (although less than CD). But it clearly can't be finite, because then EF would stand to CD in some finite proportion, but AB couldn't stand in any finite proportion to EF. So it must stand in infinite proportion to AB, and although Leibniz does not say so explicitly, in infinitely small proportion to CD as well. That means that AB is an infinitesimal, EF is infinitely larger than AB but still infinitesimal, and CD, which is finite, is infinitely larger than EF.

Next Leibniz seeks a third proportional to EF and CD; in this context, this is a line segment satisfying the equation:

$$\frac{EF}{CD} = \frac{CD}{GH}.$$

So GH is a magnitude such that CD serves as a mean proportional between EF and GH. Since EF is infinitesimal and CD is finite, GH will have to be infinitely great—if it were finite we would have a version of the same problem we had before if we took EF to be finite. Here again, Leibniz assumes the existence of such a proportion, but I will argue that this follows by an argument much like the one which guarantees the existence of the mean proportional EF. Next, Leibniz seeks for the three quantities EF, CD, and GH a fourth proportional IK satisfying:

$$\frac{EF}{CD} = \frac{CD}{GH} = \frac{GH}{IK}.$$

If, informally, we think of GH as the infinitely large quantity corresponding to EF (relative to the finite segment CD) then we may think of IK as the infinitely large quantity corresponding to the original infinitely small quantity AB.

Leibniz then extends the line GH to a line GL which has the same length as the line IK. Then the point H will lie somewhere in the middle of the line GL which has initial point G.

    G_____H_____L

    I_____K

This means, in particular, that the infinitely large quantity GH is a line segment which terminates on both ends, and Leibniz finds this absurd. This is the absurdity which contradicts the original supposition that AB was an infinitely small magnitude. Since this is a proof by contradiction, and we have derived the contradiction that an infinitely long terminating line exists, the original supposition that AB is infinitely small must be false.[3] Consequently, infinitely small magnitudes are not possible.

Leibniz's proof breaks rather naturally into two parts, and there is one main conceptual point in each. In the first part, Leibniz establishes a system of proportional relations. Here the key conceptual issue is the one we first identified in terms of the existence of the mean proportional EF. At each point along the course of the first half of the argument, Leibniz is finding more and more proportional relations. To do this he has to find new line segments standing in the right proportions. Why does he think these new line segments should exist? The second half of the proof involves embedding one of the line segments into a copy of a longer one. In this second part, Leibniz produces an infinitely long line segment

---

[3]On the status of proof by contradiction in 17th-century debates in the philosophy of mathematics, see Paolo Mancosu, *Philosophy of Mathematics and Mathematical Practice in the Seventeenth Century* (Oxford: Oxford University Press, 1996).

which terminates on both ends, which he takes to be a contradiction. The main problem in this second half of this argument is: why does Leibniz think that this is a contradiction?

Let's deal with the first problem first. Here the problem is a specific function of the way the infinite enters into the issue: if AB and CD were both finite magnitudes, there would be no problem finding a mean proportional between them. But since AB is infinitely small and CD is finite in length, we need to convince ourselves of a couple of things. The first thing we need to know is that not all infinitely small magnitudes are, so to speak, on the same level. In particular, we would be in trouble if all infinitesimals stood in some finite ratio to each other. Suppose we convince ourselves that there are infinitesimals on different levels: we still have a problem. If AB were drawn from the level of infinitesimals closest to the finite magnitudes, we would be in trouble again, because we need EF to be infinitely small but infinitely large in comparison to AB. So we need to convince ourselves not just that there are different levels of infinitesimals, but also that there is no closest level of infinitesimals to the finite magnitudes.[4]

Suppose we call a certain infinitesimal length 'dx', to signify that it represents the difference between two infinitely close values of x: $dx = x_2 - x_1$. If dx is infinitely small, then to find something infinitely small relative to it, we could try taking the difference between two infinitely close values of dx and call this ddx: $ddx = dx_2 - dx_1$. But why should these two values be ever-so-much-closer-

---

[4]These features of the situation involve the indeterminacy of first order differentials; see DLC, 24. Bos notes at 22 that all first order differentials should be at the same level, but which level is to be chosen is not fixed. Hence in the argument in the next paragraph we may choose to draw AB from the level of second order infinitesimals. My presentation is rather idealized; as Bos notes, the early practitioners of the Leibnizian calculus, including Leibniz himself, seem not to have noticed this indeterminacy of first-order differentials, 24. I treat these issues in more detail in my paper, "An enticing (im)possibility: infinitesimals, differentials, and the Leibnizian calculus," cited above.

together than x_1 and x_2?! In general, we shouldn't expect them to be, but Leibniz shows that in one particular case they are; not only that, he also shows that in this case the following proportion holds:

$$\frac{ddx}{dx} = \frac{dx}{x}. \qquad (C.1)$$

This is the content of an argument Leibniz gave in a 1695 article in the learned journal *Acta Eruditorum*, in which he responded to some arguments Bernard Nieuwentijt had made against second-order differentials, i.e. objects of the form ddx.[5] Once we have equation (C.1), essentially we may take AB to play the role of ddx and CD to play the role of x; then $EF = dx$. The details of the argument are not important here: what is important is to see the basic idea for securing the existence of the mean proportional EF.

So much for the first half of the argument. Let's turn to our question in the second half of the argument: why did Leibniz take the notion of infinitely long terminating segments to imply a contradiction? Here, our life is much easier, since Leibniz gives us a direct answer to this question in his draft of the letter to Varignon. Leibniz says, stressing that there is no need to demonstrate the existence of infinitesimals:

> ... it is unnecessary to make mathematical analysis depend on metaphysical controversies or to make sure that there are lines in nature which are infinitely small in a rigorous sense in contrast to our ordinary lines, or as a result, that there are lines infinitely greater than our ordinary ones [yet with ends; this is important inasmuch as it has seemed to me that the infinite, taken in a rigorous sense, must have its source in the unterminated;

---

[5] "Responsio ad nonnullas difficultates a Dn. Bernardo Neiwentijt circa methodum differentialem seu infinitesimalem motas," (MS, V, 320-328). A French translation with notes is provided in G.W. Leibniz, *naissance du calcul différentiel*, trans. and notes Marc Parmentier (Paris: Vrin, 1989). For commentary, see also DLC, 24.

otherwise I see no way of finding an adequate ground for distinguishing it from the finite] (PPL, 543) = (MS, IV, 91).

In the bracketed passage, which was not included in the letter as sent to Varignon, we find Leibniz making just the sort of point that he will go on to stress in the *New Essays*: it is a mistake to confuse the quantitative infinite, which has its source in the notion of the indefinite, with the idea of the infinite as a whole. The idea of infinitely long lines terminating on both ends does just that.

However, the notion of infinitely small lines terminating on both ends does, too. So why is there any need for an argument here at all? Ultimately, I think the moral of the passage from the *New Essays* we've looked at above is that there just isn't. But Leibniz needs to *work* with infinitesimals: the entire edifice of the so-called infinitesimal calculus is built upon them. Certainly the most straightforward way to understand them would be as consistent, geometric quantities. Indeed, the first half of the proof for the *impossibility* of infinitesimal quantities provides an example of just how well we *can* work with them!

Appendix D

Secondary Voices

In this Appendix I want to bring the themes from Chapter 6 into
relief by considering two book-length readings of the transition in
Wittgenstein's philosophical program. I will look first at David
Stern's *Wittgenstein of Mind and Language*, but it will be Mathieu
Marion's *Wittgenstein, Finitism, and the Foundations of Mathematics* which bears most directly on the issues of greatest concern
to me. This Appendix will be of most interest to the Wittgenstein
devotee, and others should feel free to skip over it if so inclined,
but I will do my best to make it of general relevance and interest.
To bring the long shadow of the parafinite into connection with
debates about the critical "turn" in Wittgenstein's thinking is a
focal point in the prehistory of the parafinite. It connects the main
concern of this book with the deepest strata in the thinking of
the twentieth century philosopher who thought most deeply about
issues surrounding the emergence of the mathematical parafinite.
The moral of this close-point investigation is dialectical. Wittgenstein was a radical logical conservative, emphasizing the fundamental commitment to logical determinacy above all else. Yet it was
precisely out of this deeply conservative commitment that Wittgenstein was able to bring "the long shadow of the parafinite" into view
as no other philosopher of the twentieth century. It is to the drama

of this dialectical reversal, from radical philosophical conservatism to philosophical novelty, that the investigation in this section is devoted.

D.1 David Stern's *Wittgenstein on Mind and Language*

David Stern's *Wittgenstein on Mind and Language* is, as the title suggests, focally concerned with the changes in conception of mind and language in Wittgenstein's philosophy.[1] So far as the themes of this chapter are concerned, the first thing to point out is that Stern devotes a chapter apiece to problems centering around the status of the mathematical infinite and to problems surrounding a language of immediate experience, yet he does little to tie these two sets of issues together. Consequently, although he remarks that Wittgenstein's "rejection of the notion of the completed infinite and the idea that the Russellian analysis of generality in terms of universal quantification is unsuitable for propositions concerning objects of acquaintance are closely connected with his new interest in the nature of immediate experience" (WML, 92n.8, referring then to *Philosophical Remarks* Sections X and following), he does little to tie his analyses of the problems in the two chapters together explicitly.[2]

For Stern, the transition in Wittgenstein's conceptions of mind, language and their interrelation converges on the problem of the description of immediate experience, to which Stern devotes the fifth chapter of his book. He takes Wittgenstein's general preoccupation with color language to be leading to a *replacement* of the

---

[1]David Stern, *Wittgenstein on Mind and Language* (New York: Oxford, 1995); hereafter cited internally as WML.

[2]Stern notes that the chapter on "Phenomenology" from *Philosophical Grammar* remains unpublished, WML, 94.

commitment to elementary propositions by a "primitive ontology" of "basic concepts that are supposedly constituents of every experiential judgment" (WML, 132). He then goes on to give this reconstruction of Wittgenstein's trajectory:

> Wittgenstein's abstract intuition that there must be a pictorial relationship between language and world, together with the apparent intractability of the task of specifying that relationship, had led him to the conclusion that such an analysis was unnecessary. Now [that] he had seen that he could not insist that an analysis must be possible in principle while being unable to actually give any examples, he was forced to return to the problem he had previously put to one side. (WML, 133)

This passage makes clear that Stern thinks of Wittgenstein's preoccupation with color language as "testing" the status of the language-world relationship, and specifically he views the analysis of color language as an attempt to give an example of the pictorial relationship between language and world. The "solution" at which Wittgenstein arrives is described by Stern as follows:

> By comparing the relationship between language and world with the projective geometrical relationship between two sets of plane geometrical figures whose form is clearly visible, Wittgenstein suggests that we just have to look in the right way at what is given in experience to grasp its form.

Stern's criticism of this "solution" is then that

> ...Wittgenstein had not seen the limitations of this analogy. Specifically, it had led him to expect that we should be able to identify the language that is analogous to the orthogonal projection in his analogy, the

language that directly reflects the nature of the phenomena. (WML, 134)

This reconstruction presupposes that the grammar of color is here being taken as an exemplification of *the* primary language of experience. This seems incorrect and, even further, unnecessary: even if this was Wittgenstein's intention at some point, say in "Some Remarks on Logical Form," as soon as he abandons the notion of a primary language he is no longer subject to the criticism Stern makes, since there will be *many different* languages which stand in analogy to the orthogonal projection.[3]

Consequently, it is important to distinguish carefully between the different steps along Wittgenstein's path and ask when, exactly, he took there to be a primary language of experience. Is this the case in the 1929 "Some Remarks on Logical Form"? Is it the case in the *Philosophical Remarks*? Never, sometimes, or uniformly? So far as I am able to judge, the answer to this latter question is: never. What about in the *manuscripts* of the *Philosophical Remarks* (up to a point, i.e. October 1929)? The full answer to the last of these questions must await an extensive investigation of the manuscript evidence.

In the case of the 1929 lecture "Some Remarks on Logical Form," I think an affirmative answer is still possible, but not necessary. This is possible only if we take the project of "Some Remarks on Logical Form" (already) simply to be an attempt to give the logical form of the grammar of *color* language, and this is being made analogous to a projection of one set of geometrical figures onto another. Unless we view this language as somehow "primary," Stern's "problem" does not pertain. In any case, at whatever point the rejection of a primary language of experience comes, this resolves the problem and yet leaves the possibility of the analogy to orthogo-

---

[3]There is truth to the idea that Wittgenstein will have to develop something *analogous* to the idea of projection, but *within* our language, and this role will be played by *comparison*.

nal projection intact. There is a valid concern raised by Stern's reading of the transition, but it reduces to the issue of determining when and how Wittgenstein rejected the commitment to a primary language of experience.[4]

D.2 Mathieu Marion's *Wittgenstein, Finitism, and the Foundations of Mathematics*

Marion calls Chapters 4 and 5 the "core of my reading of the transitional writings" (WFFM, xii), and so I will focus on them here. Since my criticism of Marion's reading of Wittgenstein is ultimately severe, I must emphasize at the outset that I find his work an ambitious and intelligent attempt to pursue a novel and illuminating line of intepretation. It was indeed in reading Marion's book that I first came to appreciate the value and radicalism of Wittgenstein's philosophy of mathematics, and my own reading of Wittgenstein has emerged in continuous internal conversation with his work.

D.2.1 Chapter 4: Quantification and Finitism

At the core of my concern in responding to this chapter will be the question, "In what sense is Wittgenstein a finitist?" We have already seen above that Waismann commits to a finitist position in the notes to his 1930 Königsberg lecture. Marion's claim is about Wittgenstein, however, and much of the argument for it seems to proceed in the following way. First, he shows that there are strong connection between the positions of Skolem and Wittgenstein, or Weyl and Wittgenstein, and since these other figures are finitists in

---

[4]For another view on these issues, the reader may compare Robert Alva Noë, "Wittgenstein, Phenomenology and What It Makes Sense to Say," *Philosophy and Phenomenological Research* **54** 1 (1994), 1-42.

some sense, Wittgenstein is taken to be, too. Second, Marion enlists the reactions of Wittgenstein's colleagues Russell and Carnap, who, respectively, "had already noticed [Wittgenstein's finitism] in his introduction to TLP" and in "Carnap's discussion of Wittgenstein as an ally of intuitionism [which] assumes that he accepted only [Carnap's] finitist Language I" (WFFM, 104).

The problem with all of these associations is that Marion links Wittgenstein's position with figures whom Wittgenstein clearly indicated he regarded as espousing an extensionalist logical perspective. It is indeed true that, since Wittgenstein's rejection of an extensionalist logical perspective committed him to rejecting infinite totalities in extension, he was in this regard an *ally* of a variety of finitists. But what needs stressing is that Wittgenstein rejects the conception of numbers given in extension *in toto*, and as such is no more of a "finitist" than he is an "infinitist." Indeed, this is the primary positive significance of Wittgenstein's orientation for an investigation of the parafinite. Contra Marion, I will argue that Wittgenstein's position is best characterized as logically determinist, due to his espousal of a very strong form of the principle of bivalence. Wittgenstein's criticisms of *all* these positions stem from the logical indeterminism to which they lead.

Marion effectively gives a defintion of finitism in the passage where he remarks that he will call the "account of general arithmetical propositions shared by Hilbert and Weyl, i.e. the account of them as expression of claims as opposed to statements, the 'finitist account'" (WFFM, 90). In the cases of Weyl and Hilbert this seems reasonable, given the grounds on which they stricture an account of general arithmetical propositions as statements, i.e. asserted propositions in the strict sense. Marion points out, rightly, that Weyl denies that "negations of general propositions with unrestricted quantification are not *umfangs-definit* and this is why they [i.e. general propositions] cannot be negated" and is so not lumping together universal and existential propositions in the way

Wittgenstein claims at a meeting of the Vienna Circle in January 1930 (WFFM, 96-7). This makes it clear that the criterion of strict propositionality at issue is one of *extensional* definiteness, and so in any case, as we shall see, Weyl's position must be distinguished from Wittgenstein's, where the logical criterion of definiteness is not one of *extensional* definiteness at all.

Equally, in Wittgenstein's criticism of Skolem, Marion acknowledges that Skolem does indeed transgress Wittgenstein's "idiosyncratic saying/showing distinction," and so in this regard is in no position to identify the positions of Wittgenstein and Skolem so far as sources of logical determinacy are concerned. In regard to Skolem's commitment to primitive recursive reasoning, Marion invokes W. W. Tait's characterization of the primitive recursive domain as finitary as fitting Wittgenstein's conceptions "very well," and enlists this in support of his conclusion that "[t]he resemblance [of Skolem's] with Wittgenstein's conceptions, as seen in the quotations above, is striking, and it justifies to a large extent labeling Wittgenstein as a 'finitist'" (WFFM, 99). The quotations to which Marion refers here are ones in which Wittgenstein characterizes a recursive proof as "only a general guide to an arbitrary special proof" (cited, WFFM, 98). I will return to the complicated topic of Wittgenstein's views about recursion below, but here it is important to recognize two things. First, there are indeed important commonalities between Skolem's and Wittgenstein's treatments of recursion, respectively, and it is one of the great merits of Marion's work to emphasize these. Second, however, it is largely on the basis of these commonalities that Marion associates the positions of Skolem and Wittgenstein so closely. To be fair, we should note that this is in keeping with his declared intention, stated in the Preface to his book, quoting E.M. Forster, to "only connect," and that it emerges as a natural consequence of the fact that, as he reports, his book grew out of the connections he was able to make between the views of Weyl, Ramsey and Wittgenstein on quantification in Chapter Four (WFFM, xi). However, it is then all the more important to investigate these views regarding quantification with the utmost precision.

As Marion himself admits, there are delicate issues concerning both the extreme radicalism and the instability of Wittgenstein's views about quantification circa 1929-30 which he is not able to address in Chapter Four; some of these concerns he defers to Chapter Six, and I will return to them shortly. But before doing so, it is important to get clear about those respects in which Marion sees Wittgenstein's views as radical and as unstable. The radicalism of his view is precisely what distinguishes it sharply from the view held by Weyl: namely, that it is not just mathematical propositions involving unrestricted quantification which are nonsensical, but indeed equally simple equations like '$3 + 3 = 6$' or '$25 \times 25 = 625$' which are nonsensical. Marion remarks that "[t]his radical view seems at first blush to be hardly defensible. At any rate, this view amounts to little more than the one expressed earlier in the *Tractatus Logico-Philosophicus* ..." He see this, further, as a point upon which Wittgenstein's ideas are in transition:

> ...I believe that Wittgenstein was in fact at the time in the process of changing his mind. As we shall see, he was struggling to deny that such elementary equations were devoid of 'sense'. He even introduced and discussed the idea of an analogy between mathematical equations and ordinary statements which is based on the fact that equations have some 'sense'. I shall come back to these issues in sections 6.1 and 6.2. (WFFM, 102)

We have already seen above, in the discussion of Section XV of the *Remarks*, an attempt on Wittgenstein's part to relate "impossible" to "possible" equations, and it is without a doubt that Wittgenstein was actively concerned with how mathematical equations were to be understood during the writing of the material which led to *Remarks*. However, it should be pointed out that the report by Moore which Marion cites as evidence that Wittgenstein denied that simple equations have sense comes from the period 1930-33, and so postdates the material I have been considering here. If there is an evolution away from this position, it is not effected until well beyond the material directly of concern in the period of *Philosophical Remarks*.

Indeed, in Section 6.1 Marion acknowledges that as late as 1933 Wittgenstein still held the view that, in the strictest sense, simple equations had no sense, but was willing to speak of them as having something "analogous" to sense by virtue of the fact that equations, although they cannot be proved or disproved, may be checked to hold or not to hold. But as we have seen in the commentary on Section XV from the *Remarks*, this idea of equational checking was already in place in 1929. Marion cannot have it both ways: that is, he cannot have it that in 1929 Wittgenstein was still holding a largely Tractarian conception of equations, which then underwent an evolution as a result of an instability in Wittgenstein's position. In fact, I can see no evolution in Wittgenstein's position between 1929 and 1933 on this point. Perhaps Marion's response would be that this is so much the better, and that we should simply see the radicalism of Wittgenstein's position as already moderated by an analogy between mathematical equations (strictly speaking without sense) and ordinary propositions (with sense) as early as 1929. Either way, Marion needs to acknowledge that it is something else which makes (general and particular) equations nonsensical for Wittgenstein than what makes the notion of general mathematical propositions problematic for Weyl.

Marion's strategy of reading Wittgenstein's views on the Law of Excluded Middle (LEM) is similarly compromised, for here too Marion sees Wittgenstein's views as in the process of changing in the early 1930's. Regarding LEM, Marion's interpretative strategy is outlined in the following passage:

> Wittgenstein's position could easily be construed as being incoherent. There is indeed the appearance of contradiction between the above arguments for the inapplicability of the Law of Excluded Middle to all mathematical propositions and the conclusion, reached at the end of the previous section, that the Law of Excluded Middle is valid in decidable domains. I believe that there is only an appearance of a contradiction: as I have argued, the arguments for the inapplicability of the Law of Excluded Middle to all of mathematics are rooted in the conception of mathematical equations as *Scheinsätze* in the *Tractatus Logico-Philosophicus*. In the early 1930's, Wittgenstein was simply in the process of changing his mind on this issue, introducing some legitimacy to the Law of Excluded Middle in contexts where a decision procedure is available. (WFFM, 170)

Here again Marion relies on development in Wittgenstein's position where none is to be found. Indeed, the passages that he quotes from *Remarks* and the later *Philosophical Grammar* read almost verbatim and show no development in Wittgenstein's position that I can see.

Wittgenstein's position is more radical, and in some sense simpler, than what Marion attributes to him. It is indeed the case that there are no undecidable mathematical propositions, simply because strictly speaking there are no mathematical propositions; nothing changes in this particular regard from Wittgenstein's position in the *Tractatus*, and indeed Marion's entire comparison of Wittgenstein's project to the algorithmic orientation of the lambda calculus depends on this. But, counterfactually, we may also formulate the matter in the following way: *if* there *were* mathematical propositions, there would still be no undecidable ones, because propositions must in principle have decidable truth values. So, in fact, we may say that there are no undecidable *mathematical* propositions on two different counts. When Marion is emphasizing the heritage of the logical views represented by the *Tractatus*, he even recognizes this: after emphasizing that Hilbert recognized that in the domain of ideal (infinitary) propositions it is not just the Law of Excluded Middle which does not hold, but indeed, the whole calculus of truth functions, Marion continues:

> Wittgenstein's conception is in the same vein but more radical. He argued too that his calculus of truth-functions in the *Tractatus Logico-Philosophicus* does not apply to mathematical equations. This is in line with the finitist account about general propositions involving unrestricted quantification, but it is also more radical because he extended the claim to elementary, 'real' arithmetical propositions. (WFFM, 169)

Here the talk of an "extension" is misleading, and casts Wittgenstein in the mode of a radicalizer of finitist tendencies. It should be clear by now that I draw exactly the opposite conclusion: although Wittgenstein is in certain regards a fifth column among the finitists, his *grounds* for fellow-traveling are radically distinct, *and are rooted in his radical commitment to logical determinacy.* Since this logical determinacy is reflected in his commitment to the Law of Excluded Middle applying anywhere that logic does, Marion's attempt to bring Wittgenstein into close connection with Brouwer in this regard is particularly misleading, and contravenes Wittgenstein's repeated dissociation of himself later on from the intuitionist position.[5] As we will see below, this has serious, indeed

---

[5]In "Wittgenstein and Brouwer," *Synthese* **137** 1/2 (2003), 103-127, Marion argues that Wittgenstein's position in the *Tractatus* is "quite close" to Brouwer's intuitionism, and that the residual points of disagreement lead to Wittgenstein's subsequent critique of Brouwer's intuitionism. In the course

disastrous, consequences for Marion's discussion of Wittgenstein and Strict Finitism. But before turning to these issues, we must look at Chapter Five of Marion's work, which concludes the core of his reading of Wittgenstein's "transitional writings."

D.2.2 Chapter 5: From Truth-Functional Logic to a Logic of Equations

Marion's thesis is announced in the title of his chapter: he propounds nothing less than a fundamental change in Wittgenstein's logical orientation. But where is this change in logical orientation situated? It is not the case that, according to Marion, Wittgenstein will replace truth-functional logic by a logic of equations in the general analysis of language, but rather only that Marion sees Wittgenstein adopting a new logical stance with respect to the treatment of mathematics. The immediate consequence of Marion's suggestion would be that Wittgenstein's fundamental conception of logic is bifurcated (at least), according to the domain of discourse being considered. In particular, this implies Marion's reading of Wittgenstein on the foundations of mathematics can ultimately only have peripheral significance for the overall understanding of Wittgenstein's larger philosophical program.

In fact, the reality is both more and less radical than Marion suggests. On the one hand, it is clear that Wittgenstein entertains a shift from a truth-functional logic to a logic of equations; on the other hand, the change is not one which would bifurcate logic into truth-functional and equational components. Rather, Wittgenstein

---

of his argument, Marion extends an anti-logicist reading of the *Tractatus* presented in "Qu'est-ce que l'inférence? Une relecture du *Tractatus Logico-Philosophicus*," *Archives de Philosophie* **64** (2001), 545-567. Marion's paper on Wittgenstein and Brouwer further highlights the differences between his interpretation of Wittgenstein and mine, but the roots of this difference are already apparent from my discussion of his book.

entertains a wholesale shift in logical attitude, and I will argue below that the consequences of this contemplated shift– namely the incursion of a logical indeterminacy into that very arm of Wittgenstein's enterprise which is designed to secure logical determination– are at the root of Wittgenstein's rejection of a phenomenological language. Both the contemplated shift and Wittgenstein's discomfort are registered at the very beginning of the manuscripts presented in the *Wiener Ausgabe*. In a series of passages tying the contemplated shift directly to the incursion of numbers into the domain of elementary propositions, Wittgenstein writes:

> One could certainly substitute a logic of equations for
> the logic of tautology. That is, one would pass from one
> proposition to the following through *substitution*. And
> the rules according to which these substitutions must
> be completed would be set down in equations.[6]

He goes on to ask whether propositions can be replaced by equations, so that "all of logic can appeal to rules for signs, thus in an inherited sense by way of equations,"[7] and in the next fragment draws as a summary: "I am being thrown back, against my will, to arithmetic" (WA, I, 7 entries 2, 5, 6).[8]

What, though, is Wittgenstein's motivation for entertaining the possibility of substituting an equational for a truth-functional approach? Is it motivated by a role being played by numbers in elementary propositions, or vice versa? The manuscript evidence

---

[6] "Man könnte gewiss der Logik der Tautologie eine Logik der Gleichungen setzen. D.h. man würde von einem Satz zum folgenden durch *Substitionen* gelangen. Und die Regeln nach denen diese Substitutionen vollzogen werden dürfen wären in Gleichungen niedergelegt."

[7] "die ganze Logik auf Zeichenregeln, also in einem übertragenen Sinn auf Gleichungen anwenden kann,"

[8] "Ich werde scheinbar, wider meinen Willen, auf die Arithmetik zurückgeworfen."

suggests that Wittgenstein's thoughts move *from* the consideration of perceptual and mathematical continua *to* the concern for
numbers entering into elementary propositions, but in any case,
the connection is tight and there are likely threads of implication
proceeding in both directions. I will return to these below after
completing the description of Marion's position.

The core of Marion's argument is outlined in the context of
his treatment of the problem of color exclusion. Given my contention that the considerations about color propositions do not
contribute directly to the rejection of a phenomenological language,
I would claim this already marks a limitation in Marion's perspective; how this limitation manifests itself, of course, remains to be
seen. In any case, the structure of Marion's argument is as follows: in considering color propositions Wittgenstein determined
that such propositions, and indeed "statements of degree" propositions generally, cannot be handled truth-functionally. This lack
of truth-functionality is indicated, in particular, by the fact that
numbers now enter intrinsically into such propositions and, hence,
into elementary propositions. Consequently we must find a logical
orientation which accommodates numerical propositions, and this
orientation is to be found in the logic of equations which Wittgenstein begins to develop at this time.

Several caveats must immediately be registered; for the moment, I will defer consideration of how damaging they are for Marion's position. The first is that it is not strictly speaking the case,
as Marion asserts, that any particular proposition "attributing a
given degree will have to be elementary" (WFFM, 123). What
Wittgenstein says, instead, is that statement-of-degree propositions
cannot be analysed further. By this he means that a statement-
of-degree proposition cannot be analysed into quantitative propositions plus a statement that the quantitative enumeration is exhaustive.[9] We thus cannot expect elementary propositions to respect

---

[9]Wittgenstein, "Some Remarks," 167.

truth-functional logic, but an ultimate analysis of propositions into elementary propositions is, as Wittgenstein admits in this essay, something which "has not yet been achieved,"[10] and in particular, the examples he gives of statement-of-degree propositions are "incomplete" and so not atomic. Wittgenstein does assert that numbers will enter into the atomic propositions which are required for a logical analysis of phenomena.[11] But rather than beginning with the above argument, he begins with the claim that numbers enter into atomic statements by considering "an example," noting that this is his "first definite remark on the logical analysis of actual phenomena."[12] It is indeed with reference to Wittgenstein's example that we are in a position to understand how his argument about statement-of-degree propositions proceeds several pages later in the argument.

Further, Wittgenstein gives us every reason to believe that the example he has provided is a particularly simple one and that the nature of atomic propositions, on the other hand, will be enormously complicated. The exact nature of these propositions (of which Wittgenstein's example would, presumably, supply a highly "incomplete" contribution) is puzzling enough, but even more puzzling is Wittgenstein's continued insistence, after the rejection of a phenomenological language in October 1929, that these elementary propositions (and the facts they reflect?) are highly complicated: in a discussion with the Vienna Circle on 22 December 1929 (which, indeed, Marion cites) Wittgenstein remarks:

> Just think of the equations of physics– how tremendously complex their structure is. Elementary propositions, too, will have this degree of complexity.[13]

---

[10]idem, 171.

[11]idem, 165.

[12]ibid.

[13]Ludwig Wittgenstein, *Wittgenstein and the Vienna Circle*, conversations recorded by Friedrich Waismann, ed. Brian McGuinness, trans. Joachim

That Wittgenstein would continue to talk this way in December 1929 can only mean one of three things: either that the rejection of a primary language had not completely gelled, or that the primary language Wittgenstein sought was not now a phenomenological language (but a language more akin to the language of physics), or that Wittgenstein continued to speak about elementary propositions in the absence of a primary language of any sort– but in this case, in what sense are we still speaking of *elementary* propositions?

There is clear, and early, manuscript evidence for Wittgenstein's recognition that numbers would play a role in elementary propositions: in a passage which antedates most of the material from Section XV of PR, Wittgenstein remarks:

> If my theory is correct that objects [*Gegenstände*] emerge from the manifold of real numbers in elementary propositions, this points to a more general conception of numbers– than Frege and Russell's–as I myself already had previously. I said then that number arises from the concept of calculation [*Begriff des Kalküls*], and there is certainly something to that. (WA, I, 63)[14]

This passage, however, does not distinguish between the treatment of number in the *Tractatus* and in the material from this period, but rather stresses their deep similarity.

Schulte and Brian McGuinness (New York: Barnes and Noble, 1979), 42.

[14] "Wenn meine Theorie richtig ist daß Gegenstände von der Mannigfaltigkeit der reellen Zahlen in Elementarsätzen vorkommen, so weist das auf eine allgemeinere Auggassung der Zahlen hin – als die Freges und Russells – wie ich sie selbst schon hatte. Ich sagte damals, daß die Zahl aus dem Begriff des Kalküls hervorgehe und daran ist gewiss etwas." It is interesting to note, as well, that in Wittgenstein's 22 December conversation he speaks of defects in Frege and Russell's conceptions of objects and how these conceptions are predicated on an overly close connection with subject-predicate form (*Wittgenstein and the Vienna Circle*, 41).

266

Without a fuller investigation of Wittgenstein's rejection of a phenomenological language, no firm conclusions should be drawn. However, I will suggest provisionally that we understand this situation in the following way: when Wittgenstein continues to speak of elementary propositions after the rejection of a phenomenological language, he is moving in the direction of recognizing the availability of such "elementary propositions" in *our* language and, as such, they are no longer to be identified with a (separate) primary language. The key rejection, then, is the rejection of an *independent* primary language, whether this be thought of as a phenomenological or a physical language or otherwise, and the task of finding "elementary" propositions will now be directed within our language in terms of the project of finding a "bird's-eye view" of this same language. Furthermore, there appears to be a "relativisation" of the project of linguistic analysis in the sense that we are now interested in finding "elementary" propositions with respect to a particular grammar such as, e.g., the grammar of color propositions. In this context it will be acceptable to speak of particular propositions as "elementary" despite the fact that they are far from complete, i.e. not elementary in a global sense. It does seem, however, that the recognition that we are not in any position to speak of such "globally" elementary propositions is somewhat slow in coming– how is it to be reconciled, strictly speaking, with the sort of talk we find in December 1929? Thus, to this extent, it seems that the rejection of a *primary* language is something the consequences of which do not immediately gel. In fact, this process seems in some sense to run all along the development of Wittgenstein's philosophy to the very end of his life: I would conjecture that the role played by appeal to "elementary" propositions after the rejection of a phenomenological language anticipates what Wittgenstein has to say about the logical status of certain propositions "holding fast" in the late manuscripts *On Certainty*. Hopefully, this topic will receive fuller investigation in the future.

Appendix E

Four Principles of the Parafinite

with Four Quantum Statistical Analogues

Principle of the numerical parafinite (PF 0): *Parafinite numbers are numbers that are large in the logical sense.*

Principles of quantum statistical numeration (QS 0): *Quantum statistics are numerical.*

Principle of parafinite quantification (PF 1): *The parafinite is the logically large as you please.*

Principle of quantum statistical quantification (QS 1): *Quantum statistics are collective (quantified).*

Principle of parafinite quantization (PF 2): *The parafinite is the quantitative reduction of the continuum.*

Principle of quantum statistical quantization (QS 2): *Quantum statistics are quantized (quantum).*

Principle of parafinite asymmetrization (PF 3): *The parafinite is the broken symmetry between the (de)finite and the in(de)finite.*

Principle of quantum statistical asymmetrization (QS 3): *Quantum statistics are an asymmetrization (stabilization) of classsical statistics.*

Bibliography

Adams, Robert Merrihew. *Leibniz: Determinist, Theist, Idealist* (Oxford: Oxford University Press, 1994).

Adorno, Theodor. *Alban Berg: Master of the smallest link*, trans. Juliane Brand and Christopher Hailey (Cambridge: Cambridge, 1991).

Anonymous. "Author of the Week," *The Week* July 10, 2015, 22.

Anonymous. "Bericht über die 2. Tagung für Erkenntnislehre der exakten Wissenschaften Königsberg 1930," *Erkenntnis* **2** (1931) 87-190.

Anonymous. *Number* (London: Macdonald Educational, 1970).

Aristotle. *The Complete Works of Aristotle*, 2 vols., ed. Jonathan Barnes (Princeton: Bollingen, 1984).

Badiou, Alain. *Being and Event*, trans. Oliver Feltham (London: Bloomsbury, 2013).

Baird, Joel. "Albert de Saxe et les sophismes de l'infini," in *Sophisms in Medieval Logic and Grammar*, ed. Stephen Read (Dordrecht: Kluwer, 1993), 288-303.

Barber, Richard. *The Holy Grail: Imagination and Belief* (Cambridge: Harvard, 2004).

Bassler, O. Bradley. "An enticing (im)possibility: infinitesimals, differentials, and the Leibnizian calculus," in *Infinitesimal Differences: Controversies between Leibniz and his Contemporaries*, ed. Ursula Goldenbaum and Douglas Jesseph (Berlin: Walter de

Gruyter, 2008), 135-51.

Bassler, O. Bradley. *Diagnosing Contemporary Philosophy with the Matrix Movies* (under contract, Palgrave/MacMillan).

Bassler, O. Bradley. *Kant, Shelley and the Visionary Critique of Metaphysics*, book manuscript.

Bassler, O. Bradley. *Labyrinthus de Compositione Continui: The Origins of Leibniz' Solution to the Continuum Problem*, Dissertation, University of Chicago, 1995.

Bassler, O. Bradley. "Leibniz on the Indefinite as Infinite," *The Review of Metaphysics* **51** (June 1998), 849-871.

Bassler, O. Bradley. "Miscalculation and Logical Validity," unpublished.

Bassler, O. Bradley. Review of Jaakko Hintikka's *The Principles of Mathematics Revisited* (Cambridge University Press, 1997), *The Review of Metaphysics* **51** (1997) 424-5.

Bassler, O. Bradley. Review of John P. Mayberry's *The Foundations of Mathematics in the Theory of Sets* (Cambridge: Cambridge University Press, 2000), *Notre Dame Journal of Formal Logic* **46** 1, 107-125 (2005).

Bassler, O. Bradley. Review of Mark van Atten's *On Brouwer* (Toronto: Wadsworth, 2004), *Notre Dame Journal of Formal Logic* **47** 4.

Bassler, O. Bradley. *The Pace of Modernity: Reading With Blumenberg* (Melbourne: re.press, 2012).

Bassler, O. Bradley. "The Surveyability of Mathematical Proof: A Historical Approach," *Synthese* **148** 1 (2006), 99-133.

Bassler, O. Bradley. "Towards Paris: The Growth of Leibniz's Paris Mathematics out of the Pre-Paris Metaphysics," *Studia Leibnitiana*, Band XXXI/2 (1999), 159-180.

Becker, Oskar. *Dasein und Dawesen: Gesammelte philosophische Aufsätze* (Pfullingen: Neske, 1963).

Bernays, Paul. "The Philosophy of Mathematics and Hilbert's Proof Theory," translated in Paolo Mancosu, *From Brouwer to*

Hilbert: The Debatge on the Foundations of Mathematics in the 1920's, 234-265.

Bernays, Paul. "On Platonism in Mathematics," reprinted in *Philosophy of Mathematics*, 2nd ed., ed. P. Bennaceraf and H. Putnam (Cambridge: Cambridge University Press, 1983), 258-71.

Bishop, John. *Joyce's Book of the Dark,* **Finnegans Wake** (Madison: Wisconsin, 1986).

Blumenberg, Hans. *Beschreibung des Menschen*, ed. Manfred Sommer (Frankfurt: Suhrkamp, 2006).

Blumenberg, Hans. *The Legitimacy of the Modern Age*, trans. Robert Wallace (Cambridge: MIT, 1983).

Blumenberg, Hans. *Ein mögliches Selbstverständnis* (Stuttgart: Reclam, 1997).

Boolos, George, and Jeffrey, Richard. *Computability and logic* (Cambridge: Cambridge University Press, 1974).

Borel, Emile. *Valeur pratique et philosophie des probabilités* (Paris: Gauthier-Villars, 1937).

Bos, H. J. M. "Differentials, "Higher-Order Differentials and the Derivative in the Leibnizian Calculus," *Archive for History of Exact Science* **14**, 1-90.

Brakhage, Stan. *By Brakhage: An Anthology*, 2 vols., CC1590D and CC1898D (The Criterion Collection, 2003, 2010).

Burke, Edmund. *Philosophical Inquiry into the Origin of Our Ideas of the Sublime and Beautiful* (1757; 6th ed., 1770).

Burton, Robert. *The Anatomy of Melancholy*, ed. Floyd Dell and Paul Jordan-Smith (New York: Tudor Publishing Company, 1927).

Carroll, Lewis. (Charles Dodgson) *The Hunting of the Snark*, in *The Humorous Verse of Lewis Carroll* (repr. New York: Dover Publications, 1960)

Cavell, Stanley. *A Pitch of Philosophy: Autobiographical Exercises* (Cambridge: Harvard, 1994).

Clark, T. J. "Pollock's Smallness," in *Jackson Pollock: New*

Approaches, ed. Kirk Varnedoe and Pepe Karmel (New York: Museum of Modern Art, 1999), 15-31.

Clavelin, Maurice. *The Natural Philosophy of Galileo: Essay on the Origins and Formation of Classical Mechanics*, trans. A. J. Pomerans (Cambridge: MIT, 1974).

Clavelin, Maurice. "Le Problème du Continu et les Paradoxes de l'Infini chez Galilée," *Thales* 10 (1959), 1-26.

Coffa, J. Alberto. *The Semantic Tradition from Kant to Carnap: To the Vienna Station*, ed. Linda Wessels (Cambridge: Cambridge, 1991).

Costabel, Pierre. "De Scientia Infinitii," in *Leibniz, aspects de l'homme et de l'oeuvre* (Paris, 1966).

Crosby, Alfred W. *Ecological Imperialism: The Biological Expansion of Europe, 900-1900* (Cambridge: Cambridge, 1986).

Dantzig, D. van. "Is $10^{10^{10}}$ a Finite Number?," *Dialectica* 9 (1956) 273-7, reprinted in Richard L. Epstein and Walter A. Carnielli, *Computability: Computable Functions, Logic, and the Foundations of Mathematics*, 260-3.

Dawson, John. "The Reception of Gödel's Incompleteness Theorem," in *Gödel's Theorem in Focus*, 74-95.

Dedekind, Richard. *Essay on the Theory of Numbers*, trans. Wooster Woodruff Beman (repr. New York: Dover, 1963).

Deleuze, Gilles. *Difference and Repetition*, trans. Paul Patton (New York: Columbia, 1994).

de l'Hospital, G. F. A. *Analyse des infiniments petites pour l'intelligence des lignes courbes* (Paris: 1st ed. 1696, 2nd ed. 1715).

Descartes, René, *Oeuvres de Descartes*, ed. Adam and Tannery, new presentation in 12 vols. (Paris: Vrin, 1964-76).

Descartes, René, *The Philosophical Writings of Descartes*, 3 vols., trans. Cottingham, Stoothoff, Murdoch and Kenny (Cambridge: Cambridge University Press, 1985-1991).

Dijksterhuis, E. J. *Archimedes*, trans. C. Dikshoorn, with a new bibliographic essay by Wilbur R. Knorr (Princeton: Princeton,

1987).

Drake, Frank R. *Set Theory: An Introduction to Large Cardinals* (Amsterdam: North-Holland, 1974.

Duchamp, Marcel. *Marcel Duchamp*, ed. Anne D'Harnoncourt and Kynaston McShine (Greenwich: MOMA/New York Graphic Society, 1973).

Dummett, Michael. *Truth and Other Enigmas* (Cambridge: Harvard, 1978).

Eames, Charles and Eames, Ray. Volume I of *The Films of Charles and Ray Eames* ID921OPIDVD (Lucia Eames DBA Eames Office, 1989), distributed exclusively by Image Entertainment.

Eley, Lothar. *Metakritik der Formalen Logik: Sinnliche Gewissheit als Horizont der Aussagenlogik und Elementaren Prädikatenlogik* (The Hague: Martinus Nijhoff, 1969).

Enderton, Herbert B. *A Mathematical Introduction to Logic* (Orlando: Academic Press, 1972).

Epstein, Richard L., and Carnielli, Walter A. *Computability: Computable Functions, Logic and the Foundations of Mathematics*, 2nd edition (Belmont: Wadsworth, 2000).

Feldman, Morton. *Give My Regards To Eight Street: Collected Writings of Morton Feldman*, ed. B. H. Friedman, afterword by Frank O'Hara (Cambridge: Exact Change, 2000).

Field, Richard S. *Mel Bochner: Thought Made Visible 1966-1973* (New Haven: Yale University Art Gallery, 1995).

Finkelstein, D. R. *Quantum Relativity: A Synthesis of the Ideas of Einstein and Heisenberg* (Berlin: Springer, 1996).

Fowler, David. *The Mathematics of Plato's Academy: A New Reconstruction*, 2nd ed. (Oxford: Clarendon Press, 1999).

Frege, Gottlob. *The Foundations of Arithmetic*, German Text with English trans. J. L. Austin (Evanston: Northwestern, 1978).

Galilei, Galileo. *Dialogues Concerning Two New Sciences*, trans. Henry Crew (New York: MacMillan, 1914, repr. Dover and also Northwestern University Press).

Galilei, Galileo. *Opere di Galileo Galilei* (Florence: Edizione Nazionale, 1890-1910), vol. 8.

Galilei, Galileo. *Two New Sciences*, trans. Stillman Drake (Madison: University of Wisconsin Press, 1974).

Ganea, Mihai. *Arithmetic without Numbers*, Dissertation, University of Illinois (Chicago), 1995.

Girard, Jean-Yves. *The Blind Spot: Lectures on Logic* (Zürich: European Mathematical Society, 2011).

Gödel, Kurt. *Collected Works*, vol. 1: *Publications 1929-1936*, ed. Feferman et al. (New York: Oxford Press, 1986).

Gödel, Kurt. "Die Vollständigkeit der Axiome des logischen Funktionenkalküls," *Monatshefte für Mathematik und Physik* **37** (1930), 349-360, reprinted with face-à-face translation by Stefan Bauer-Mengelberg in Kurt Gödel, *Collected Works*, vol. 1, 102-123.

Gold, Thomas. *The Nature of Time* (Ithaca: Cornell, 1967).

Grattan-Guinness, I. Review of David Fowler, *The Mathematics of Plato's Academy: A New Reconstruction*, in *Mathematical Gazette* **84** 499, 165-6.

Gutzwiller, Martin. C. *Chaos in Classical and Quantum Mechanics* (New York: Springer, 1990).

Hardy, G. H. *Divergent Series* (Oxford: Clarendon, 1949).

Heller, Michael, and Woodin, W. Hugh., eds., *Infinity: New Research Frontiers* (Cambridge: Cambridge, 2011).

Heidegger, Martin. *Basic Questions of Philosophy: Selected "Problems" of "Logic,"* trans. Richard Rojcewicz and André Schuwer (Bloomington: Indiana, 1994).

Hintikka, Jaakko. *The Principles of Mathematics Revisited* (Cambridge: Cambridge University Press, 1996).

Hodges, Wilfrid. *Building models by games* (Cambridge: Cambridge University Press, 1985).

Hofmann, Josef. E. *Leibniz in Paris 1672-1676: His Growth to Mathematical Maturity* (Cambridge: Cambridge University Press,

1974.

Hopkins, Burt C. *The Origin of the Logic of Symbolic Mathematics: Edmund Husserl and Jacob Klein* (Bloomington: Indiana, 2011).

Husserl, Edmund. *Introduction to the Logical Investigations*, trans. Philip J. Bossert and Curtis H. Peters (The Hague: Martinus Nijhoff, 1975).

Husserl, Edmund. *Philosophie der Arithmetik, mit ergänzenden texten (1890-1901)*, ed. Lothar Eley (The Hague: Martinus Nijhoff, 1970).

Husserl, Edmund. *Philosophy of Arithmetic: Psychological and Logical Investigations, with Supplementary Texts from 1887-1901*, trans. Dallas Willard (Dordrecht: Kluwer, 2003).

Isles, David. "Remarks on the Notion of Standard Non-Isomorphic Natural Number Series," in *Constructive Mathematics: Proceedings, New Mexico, 1980*, Springer Lecture Notes in Mathematics vol. 873 (Berlin: Springer Verlag, 1981), 111-34; partially reprinted in Epstein and Carnielli, *Computability*, pp. 263-70.

Isles, David. "Theorems of Peano Arithmetic are Buridan-Volpin Recursively Satisfiable," *Reports on Mathematical Logic* **31** (1997), 57-74.

Isles, David. "What Evidence is There That 2^{65536} is a Natural Number?," *Notre Dame Journal of Formal Logic* **33** 4 (1992), 465-80.

James, Carol P. "Duchamp's Silent Noise / Music for the Deaf," in *Marcel Duchamp: Artist of the Century*, ed. Rudolf E. Kuenzli and Francis M. Naumann (Cambridge: MIT, 1989) 106-26.

Joyce, James. *Finnegans Wake* (New York: Viking, 1939).

Joyce, James. *Ulysses* (New York: Vintage, 1986).

Kant, Immanuel. *Critique of Pure Reason*, trans. Werner Pluhar with an intro. by Patricia Kitcher (Indianapolis: Hackett, 1996).

Klein, Jacob. *Greek Mathematical Thought and the Origin of*

Algebra, trans. Eva Brann (repr. New York: Dover, 1992).

Knobloch, Eberhard. "Galileo and Leibniz' Different Approaches to Infinity," *Archive for the History of Exact Sciences* **54** (1999), 87-99.

Knorr, W. R. *The Evolution of the Euclidean Elements* (Dordrecht: Reidel, 1975).

Koyré, Alexandre. *From the Closed World to the Infinite Universe* (Baltimore: Johns Hopkins University Press, 1957).

Kreisel, G., Mints, G. E., and Simpson, S. G. "The Use of Abstract Language in Elementary Metamathematics: Some Pedagogic Examples," *Logic Colloquium*, Springer Lecture Notes in Mathematics 453, 38-131.

Lakatos, Imre. "Cauchy and the continuum: the significance of non-standard analysis for the history and philosophy of mathematics," ed. J.P. Cleave, in Imre Lakatos, *Philosophical Papers Volume 2*, 43-60.

Lakatos, Imre. *Philosophical Papers Volume 2: Mathematics, science and epistemology*, ed. John Worrall and Gregory Currie (Cambridge: Cambridge University Press, 1978).

Lambek, J. and Scott, P. J. *Introduction to higher order categorical logic* (Cambridge: Cambridge University Press, 1986).

Lapidus, M. L. and Frankenhuysen, M. van. *Fractal Geometry, Complex Dimensions and Zeta Functions: Geometry and spectra of fractal strings* (New York: Springer, 2006).

Lapidus, M. L. *In Search of the Riemann Zeros: Strings, Fractal Membranes and Noncommutative Spacetimes* (Providence: AMS, 2008).

Lavine, Shaughan. *Understanding the Infinite* (Cambridge: Harvard, 1994).

Leibniz, Gottfried Wilhelm. *De l'Horizon de la Doctrine Humaine: Ἀποκατάστασις πάντων (La Restitution universelle) (1715)*, trans. and annotated Michel Fichant (Paris: Vrin, 1991).

Leibniz, Gottfried Wilhelm. *De quadratura arithmetica circuli ellipseos et hyperbolae cujus corollarium est trigonometria sine tabulis*, ed. Eberhard Knobloch (Göttingen, Vandenhoeck & Ruprecht, 1993).

Leibniz, Gottfried Wilhelm. *De Summa Rerum: Metaphysical Papers, 1675-1676*, trans. G.H.R. Parkinson (New Haven: Yale University Press, 1992).

Leibniz, Gottfried Wilhelm. *Mathematische Schriften*, ed. C.I. Gerhardt (Halle, 1849-63, repr. Olms Verlag).

Leibniz, Gottfried Wilhelm. *naissance du calcul différentiel*, trans. and notes Marc Parmentier (Paris: Vrin, 1989).

Leibniz, Gottfried Wilhelm. *New Essays on Human Understanding*, trans. and ed. Peter Remnant and Jonathan Bennett (Cambridge: Cambridge University Press, 1981).

Leibniz, Gottfried Wilhelm. *Philosophical Essays*, trans. Ariew and Garber (Indianapolis: Hackett, 1989).

Leibniz, Gottfried Wilhelm. *Philosophical Papers and Letters*, trans. and ed. Leroy Loemker (Dordrecht: Reidel, 1969).

Leibniz, Gottfried Wilhelm. *Die Philosophische Schriften von Gottfried Wilhelm Leibniz*, ed. C.I. Gerhardt (Berlin 1875-90, repr. Olms Verlag).

Leibniz, Gottfried Wilhelm. *quadrature arithmétique du cercle, de l'ellipse et de l'hyperbole*, ed. Eberhard Knobloch, trans. with introduction and notes by Marc Parmentier (Paris: Vrin, 2004).

Leibniz, Gottfried Wilhelm. *Sämtliche Schriften und Briefe* (Darmstadt/Leipzig/Berlin, 1923-).

Lévi-Strauss, Claude. *The Savage Mind*, trans. from the French (Chicago: Chicago, 1966).

Lorenzen, Paul. *Constructive philosophy*, trans. Karl Richard Pavlovic (Amherst: University of Massachusetts Press, 1987).

Mahnke, Dietrich. "Die Enstehung des Funktionsbegriffes," *Kantstudien* **31** (1926), 426-28.

Mancosu, Paolo, *From Brouwer to Hilbert: The Debate on the Foundations of Mathematics in the 1920s* (New York: Oxford,

279

1998).

Mancosu, Paolo. *Philosophy of Mathematics and Mathematical Practice in the Seventeenth Century* (Oxford: Oxford University Press, 1996).

Mannoury, Gerritt. *Methodologisches und Philosophisches zur Elementarmathematik* (Haarlem: P. Visster, 1909).

Mannoury, Gerritt. *Woord en Gedachte*, (Groningen, 1931).

Marion, Mathieu. "Qu'est-ce que l'inférence? Une relecture du *Tractatus Logico-Philosophicus*," *Archives de Philosophie* **64** (2001), 545-567.

Marion, Mathieu. "Wittgenstein and Brouwer," *Synthese* **137** 1/2 (2003), 103-127.

Marion, Mathieu. *Wittgenstein, Finitism, and the Foundations of Mathematics* (Oxford: Clarendon Press, 1998).

Mayberry, J. P. *The Foundations of Mathematics in the Theory of Sets* (Cambridge: Cambridge, 2000).

McCauley, J. L. *Chaos, Dynamics and Fractals: an algorithmic approach to deterministic chaos* (Cambridge: Cambridge, 1993).

Meli, Domenico Bertoloni. *Equivalence and Priority: Newton versus Leibniz, including Leibniz's Unpublished Manuscripts on the Principia* (Oxford: Clarendon, 1993).

Miller, Mitchell. "Figure, Ratio, Form: Plato's Five Mathematical Studies," preprint.

Morrison, Philip, and Morrison, Phylis, and the office of Charles and Ray Eames. *Powers of Ten: About the Relative Size of the Universe* (New York: Scientific American Books, 1982).

Mostowski, Andrzej. *Thirty Years of Foundational Studies: Lectures on the Development of Mathematical Logic and the Study of The Foundations of Mathematics in 1930-1964* (New York: Barnes and Noble, 1966).

Mueller, Ian. *Philosophy of Mathematics and Deductive Structure in Euclid's Elements* (Cambridge, MA), 1981.

Mycielski, Jan. "Analysis Without Actual Infinity," *Journal of*

Symbolic Logic **46** (1981), 625-33.

Mycielski, Jan. "The Meaning of Pure Mathematics," *Journal of Philosophical Logic* **18** (1989) 315-20.

Nakamura, Katsuhiro. *Quantum Chaos: A New Paradigm of Nonlinear Dynamics* (Cambridge: Cambridge, 1993).

Nelson, Edward. *Predicative Arithmetic* (Princeton: Princeton, 1986).

Nelson, Edward. "Mathematical Mythologies," in *Le Laybrinthe du Continu*, ed. Jean-Michel Salanskis and Hourya Sinaceur (Paris: Springer, 1992), 155-67.

Nelson, Edward. "Warning Signs of a Possible Collapse of Contemporary Mathematics," in Michael Heller and W. Hugh Woodin, eds., *Infinity: New Research Frontiers* (Cambridge: Cambridge, 2011), 76-85.

Netz, Reviel. *Ludic Proof: Greek Mathematics and the Alexandrian Aesthetic* (Cambridge: Cambridge, 2009).

Netz, Reviel. *The Shaping of Deduction in Greek Mathematics: A Study in Cognitive History* (Cambridge: Cambridge, 1999).

Netz, Reviel. *The Transformation of Mathematics in the Early Mediterranean World: From Problems to Equations* (Cambridge: Cambridge, 2004).

Noë, Robert Alva, "Wittgenstein, Phenomenology and What It Makes Sense to Say," *Philosophy and Phenomenological Research* **54** 1 (1994), 1-42.

Ockham, William. *Philosophical Writings: A Selection*, trans. Philotheus Boehner, O.F.M., revised by Stephen Brown (Indianapolis: Hackett, 1990).

Omnes, Roland. *Understanding Quantum Mechanics* (Princeton: Princeton, 1999).

Peirce, C. S. "How to Make Our Ideas Clear," in *The Essential Peirce: Selected Philosophical Writings*, vol. 1 (1867-1893), ed. Nathan Houser and Christian Kloesel (Bloomington: Indiana, 1992), 124-41

Peirce, C. S. *Reasoning and the Logic of Things* (Cambridge: Harvard University Press, 1992).

Plato. *Republic*, trans. G.M.A. Grube revised by C.D.C. Reeve (Indianapolis: Hackett, 1992).

Poe, Edgar Allan. *Poetry, Tales, & Selected Essays* (New York: Library of America College Editions, 1996).

Proust, Marcel, *Remembrance of Things Past*, trans. C. K. Scott Moncrieff and Terence Kilmartin; and by Andreas Mayor (New York: Random House, 1981).

Ramsey, F. P. *The Foundations of Mathematics and Other Logical Essays*, ed. R. B. Braithwaite (London: Routledge, 1931).

Richards, I. A. *The Philosophy of Rhetoric* (Oxford: Oxford, 1936).

Rieger, L. "Sur le problème des nombres naturels," in *Infinitisitic Methods (Proceedings of the Symposium of the Foundations of Mathematics, Warsaw, 1959)* (Oxford: Pergamon Press, 1961), 225-33.

Robinson, Abraham. "The Metaphysics of the Calculus," in *The Philosophy of Mathematics*, ed. Jaakko Hintikka (Oxford: Oxford, 1969), 153-63.

Robinson, Abraham. *Non Standard Analysis* (Amsterdam: North Holland, 1966), repr. Princeton, 1996.

Russell, Bertrand. *Foundations of Logic 1903-1905: The Collected Papers of Bertrand Russell*, vol. 4 (London: Routledge, 1994).

Russell, Bertrand. "On the Relations of Number and Quantity," Mind n. s. **6** (July 1897): 326-41, reprinted in Bertrand Russell, *Philosophical Papers 1896-99: The Collected Writings of Bertrand Russell*, vol. 2 (London: Unwin Hyman, 1990).

Russell, Bertrand. *Principles of Mathematics* (repr. New York: Norton, 1964).

Russell, Bertrand. *Toward the Principles of Mathematics 1900-1902: The Collected Papers of Bertrand Russell*, vol. 3 (London:

Routledge, 1993).

Schmit, Roger. *Husserls Philosophie der Mathematik: Platonische und konstruktivistische Momente in Husserls Mathematikbegriff* (Bonn: Bouvier, 1981).

Schulman, L. S. *Time's arrows and quantum measurement* (Cambridge: Cambridge, 1997).

Salanskis, Jean-Michel. *Le Constructivisme non-standard* (Paris: Presses Universitaires du Septentrion, 1999).

Segre, Michael. *In the Wake of Galileo*, with a forward by I. Bernard Cohen (New Brunswick: Rutgers University Press, 1991).

Settle, Mary Lee. *The Story of Flight*, illus. George Evans (New York: Random House, 1967).

Shanker, S. G., ed. *Gödel's Theorem in Focus* (New York: Routledge, 1988).

Shapiro, Stewart. *Foundations without Foundationalism* (New York: Oxford University Press, 1991).

Shea, William R. "Galileo's Atomic Hypothesis," *Ambix* **17** (1970), 13-27.

Shields, Philip R. *Logic and Sin in the Writings of Ludwig Wittgenstein* (Chicago: Chicago, 1993).

Smith, A. Mark. "Galileo's Theory of Indivisibles: Revolution or Compromise?," *Journal of the History of Ideas* **37** (1976), 571-88.

Stein, Howard. "Eudoxus and Dedekind: On the Ancient Greek Theory of Ratios and its Relation to Modern Mathematics," *Synthese* **84** (1990) 163-211.

Steiner, George. *In Bluebeard's Castle: Some Notes Towards a Redefinition of Culture* (New Haven: Yale, 1971).

Stern, David. *Wittgenstein on Mind and Language* (New York: Oxford, 1995).

Tiles, Mary. *The Philosophy of Set Theory* (repr. Mineola: Dover, 2004).

Vopěnka, Petr. *Mathematics in the Alternative Set Theory*

(Leipzig: Teubner Verlagsgesellschaft, 1979).

Vopěnka, Petr, and Hájek, Petr. *The Theory of Semisets* (Amsterdam: North-Holland, 1972).

Waismann, Friedrich. *Lectures on the Philosophy of Mathematics*, ed. Wolfgang Grassl (Amsterdam: Rodopi, 1982).

Waismann, Friedrich. "The Nature of Mathematics: Wittgenstein's Standpoint," trans. S. G. Shanker, in *Ludwig Wittgenstein: Critical Assessments*, Volume Three: *From the **Tractatus** to **Remarks on the Foundations of Mathematics***, 60-67.

Weissert, Thomas P. *The Genesis of Simulation in Dynamics: Pursuing the Fermi-Pasta-Ulam Problem* (New York: Springer, 1997).

Weyl, Hermann. *The Continuum: A Critical Examination of the Foundations of Analysis*, trans. Stephen Pollard and Thomas Bole with a Foreword by John Archibald Wheeler (repr. New York: Dover, 1994).

Wittgenstein, Ludwig. *On Certainty*, ed. G. E. M. Anscombe and G. H. von Wright, trans. Denis Paul and G. E. M. Anscombe (New York: Harper & Row, 1969).

Wittgenstein, Ludwig. *Philosophical Grammar*, trans. A. Kenny (Berkeley: California, 1974).

Wittgenstein, Ludwig. *Philosophical Remarks*, ed. Rush Rhees, trans. Raymond Hargreaves and Roger White (Chicago: Chicago, 1975).

Wittgenstein, Ludwig. "Some Remarks on Logical Form," *Proceedings of the Aristotelian Society,* Supplementary Volume 9 (1929), 162-71.

Wittgenstein, Ludwig. *Tractatus-Logico-Philosophicus*, trans. D. F. Pears and B. F. McGuinness (London: Routledge, 1974).

Wittgenstein, Ludwig. *Wiener Ausgabe: Studien Texte*, vols. 1-5, ed. Michael Nedo (repr. Frankfurth am Main: Zweitausendeins, 1994-1996).

Wittgenstein, Ludwig. *Wittgenstein and the Vienna Circle,*

conversations recorded by Friedrich Waismann, ed. Brian McGuinness, trans. Joachim Schulte and Brian McGuinness (New York: Barnes and Noble, 1979).

Wittgenstein, Ludwig. *Wittgenstein's Lectures, Cambridge, 1930-1932*, ed. D. Lee (Totowa, N. J.: Rowman and Littlefield, 1980).

Wittgenstein, Ludwig. *Wittgenstein's Lectures on the Foundations of Mathematics, Cambridge 1939*, ed. Cora Diamond (Ithaca: Cornell, 1976).

Yessenin-Volpin, A. S. "The ultra-intuitionistic criticism and the antitraditional program for foundations of mathematics," in A. Kino, J. Myhill, and R. E. Vesley, *Intuitionism and Proof Theory* (Amsterdam: North-Holland, 1970), 3-45.

Zalamea, Fernando. *Peirce's Logic of Continuity* (Boston: Docent, 2012).

Zalamea, Fernando. *Synthetic Philosophy of Contemporary Mathematics* (trans. Zachary Luke Fraser, New York: Sequence Press, original publication in Spanish 2009).

www.ingramcontent.com/pod-product-compliance
Lightning Source LLC
Chambersburg PA
CBHW060323200326
41519CB00011BA/1814